Fabrizio Zanta

Folding and Self-Association Studies of Id3 Dimerization Domain

Fabrizio Zanta

Folding and Self-Association Studies of Id3 Dimerization Domain

Südwestdeutscher Verlag für Hochschulschriften

Impressum / Imprint

Bibliografische Information der Deutschen Nationalbibliothek: Die Deutsche Nationalbibliothek verzeichnet diese Publikation in der Deutschen Nationalbibliografie; detaillierte bibliografische Daten sind im Internet über http://dnb.d-nb.de abrufbar.
Alle in diesem Buch genannten Marken und Produktnamen unterliegen warenzeichen-, marken- oder patentrechtlichem Schutz bzw. sind Warenzeichen oder eingetragene Warenzeichen der jeweiligen Inhaber. Die Wiedergabe von Marken, Produktnamen, Gebrauchsnamen, Handelsnamen, Warenbezeichnungen u.s.w. in diesem Werk berechtigt auch ohne besondere Kennzeichnung nicht zu der Annahme, dass solche Namen im Sinne der Warenzeichen- und Markenschutzgesetzgebung als frei zu betrachten wären und daher von jedermann benutzt werden dürften.

Bibliographic information published by the Deutsche Nationalbibliothek: The Deutsche Nationalbibliothek lists this publication in the Deutsche Nationalbibliografie; detailed bibliographic data are available in the Internet at http://dnb.d-nb.de.
Any brand names and product names mentioned in this book are subject to trademark, brand or patent protection and are trademarks or registered trademarks of their respective holders. The use of brand names, product names, common names, trade names, product descriptions etc. even without a particular marking in this work is in no way to be construed to mean that such names may be regarded as unrestricted in respect of trademark and brand protection legislation and could thus be used by anyone.

Coverbild / Cover image: www.ingimage.com

Verlag / Publisher:
Südwestdeutscher Verlag für Hochschulschriften
ist ein Imprint der / is a trademark of
OmniScriptum GmbH & Co. KG
Heinrich-Böcking-Str. 6-8, 66121 Saarbrücken, Deutschland / Germany
Email: info@svh-verlag.de

Herstellung: siehe letzte Seite /
Printed at: see last page
ISBN: 978-3-8381-3984-5

Zugl. / Approved by: Bochum, Ruhr-Universität Bochum, Dissertation, 2013

Copyright © 2014 OmniScriptum GmbH & Co. KG
Alle Rechte vorbehalten. / All rights reserved. Saarbrücken 2014

Contents

List of Abbreviations	**5**
1 Id proteins - Biology, structure and modulation	**9**
1.1 The HLH protein family and the Id proteins	9
1.2 Mode of action and regulation of the Id proteins	11
1.3 Structural studies of the Id HLH dimerization domain	12
1.4 Synthesis and conformational analysis of Id2 protein fragments	15
1.5 Peptidomimetics as potential modulators of protein-protein interactions	16
1.6 Current structural information of Id3	17
1.7 Scope and objectives	20
References	**21**
2 Design and synthesis of Id3 protein fragments for conformational studies	**25**
2.1 Introduction	25
2.2 Solid-phase synthesis of the Id3 protein fragments	26
2.3 Conclusions	32
References	**33**
3 Covalent dimers of the Id3 HLH domain	**35**
3.1 Introduction	35
3.2 Peptide design	36
3.3 Results	39
3.3.1 Formation of the oxidized homodimers of Id3 HLH analogs	39

		3.3.2 Disulfide reshuffling and thiol-disulfide exchange assays	40
	3.4	Discussion	42
	3.5	Conclusions	44

References 47

4 Studies of the self-association of oligo-Arg/Glu-tagged analogs of the Id3 HLH domain 49

- 4.1 Introduction 49
- 4.2 Peptide synthesis 50
- 4.3 Results and discussion 51
  - 4.3.1 Conformational studies by CD spectroscopy 51
  - 4.3.2 Thermal dissociation of the helix bundle 56
- 4.4 Conclusions 58

References 61

5 A FRET study on self-association of Id3 protein fragments 63

- 5.1 Introduction 63
- 5.2 Design and solid-phase synthesis of the fluorescently labeled Id3 protein fragments 64
- 5.3 Results 67
  - 5.3.1 Conformational studies by CD spectroscopy 67
  - 5.3.2 Fluorescence spectroscopy 69
  - 5.3.3 Self-recognition between partially and fully folded HLH domains 74
- 5.4 Conclusions 76

References 77

6 Salt and solvent effects on the conformational stability of the Id HLH domain 79

- 6.1 Introduction 79
- 6.2 Results 81

- 6.2.1 Chol-dhp negatively affects the helical content of the Id3 HLH domain 81
- 6.2.2 Chol-Cl positively affects the helix bundles of the Id3 HLH domain 83
- 6.2.3 Chol-dhp favors tight and highly structured helix bundles of the Id4 HLH domain 83
- 6.2.4 Chol-Cl negatively affects the helical content of the Id4 HLH domain 86
- 6.2.5 Effect of chol-dhp and chol-Cl on the thermal dissociation of the helix bundles 86
- 6.2.6 Chol-dhp-stabilized helix bundles are resistant to the denaturing effect of GndCl 88
- 6.2.7 Chol-Cl strengthens the denaturing effect of GndCl on the helix bundles 90
- 6.2.8 Chol-dhp supports the TFE-induced unbundling of the helix bundles 90
- 6.3 Discussion 92
- 6.4 Conclusions 96

References 99

7 Experimental part 103
- 7.1 Materials 103
- 7.2 Methods 104
  - 7.2.1 Solid phase peptide synthesis (SPPS) 104
  - 7.2.2 Circular dichroism (CD) spectroscopy [6] 107
  - 7.2.3 Steady-state fluorescence spectroscopy 108
- 7.3 Standard procedure for polypeptide synthesis 109
  - 7.3.1 Peptide chain assembly by automated SPPS 109
  - 7.3.2 Peptide chain elongation by manual SPPS 110
  - 7.3.3 Labeling of the Id3 fragments 110
- 7.4 General procedure for peptide purification and characterization 111
- 7.5 Mass spectrometry 111
- 7.6 UV and CD spectroscopy 112

	7.6.1 Procedure for thermal denaturation by CD spectroscopy	112
7.7	Oxidation experiments	113
	7.7.1 Monomer oxidation	113
	7.7.2 Disulfide reshuffling and thiol-disulfide exchange measurements	113
7.8	Procedure for fluorescence spectroscopy analysis	113
	7.8.1 Fluorescence spectroscopy of the labeled Id3 peptides	113
	7.8.2 Titration of partially folded Id3 HLH with folded Id3 HLH	114

References   115

Summary   117

Appendix   121

Acknowledgements   131

List of abbreviations

Amino acid codes

Amino acid	One letter code	Three letter code
Alanine	A	Ala
Cysteine	C	Cys
Aspartic acid	D	Asp
Glutamic acid	E	Glu
Glycine	G	Gly
Histidine	H	His
Isoleucine	I	Ile
Lysine	K	Lys
Leucine	L	Leu
Methionine	M	Met
Asparagine	N	Asn
Proline	P	Pro
Glutamine	Q	Gln
Arginine	R	Arg
Serine	S	Ser
Threonine	T	Thr
Valine	V	Val
Tyrosine	Y	Tyr

Abbreviation	Extended name
Ac	acetyl
ACN	acetonitrile
bHLH	basic helix-loop-helix
Boc	*tert*-butyloxycarbonil
CD	circular dichroism
chol	choline
DCM	dichloromethane
dhp	dihydrophosphate
DIC	N,N'-diisopropylcarbodiimide
DIPEA	diisopropilethylamine
DMF	dimethylformamide
EDT	ethanedithiol
FAM	carboxyfluorescein
Fmoc	fluorenyl-9-methoxycarbonyl
FRET	fluorescence resonance energy transfer
HBTU	O-(1-benzotriazolyl)-N,N,N',N'-tetramethyluroniumhexafluoro-phosphate
HLH	helix-loop-helix
HOBt	hydroxybenzotriazole
Id	inhibitor of DNA binding or inhibitor of differentiation
IL	ionic liquid
LZ	leucine zipper
MBHA	4-methylbenzhydrylamine
MALDI-TOF	Matrix-Assisted-Laser-Desorption-Ionization Time of Flight
MS	Mass spectrometry
MMP	matrix metalloproteinase
Mtt	4-methyltrityl
MW	Molecular Weight
MyoD	myogenic determination factor
NES	nuclear export signal

Abbreviation	Extended name
NMP	N-methyl-2-pyrrolidone
NMR	nuclear magnetic resonance
PG	protecting group
RMSD	root-mean-square deviation
RP-HPLC	reverse phase high performance liquid chromatography
R.T.	room temperature
SPPS	solid phase peptide synthesis
TAMRA	tetramethylrhodamine
TFA	trifluoroacetic acid
TFE	2,2,2-trifluoroethanol
TIS	triisopropylsilane
T_m	melting temperature
TMSBr	trimethylsilylbromide
UV	ultra violet

Chapter 1

Id proteins - Biology, structure and modulation

1.1 The HLH protein family and the Id proteins

The transcription of DNA to RNA is controlled by proteins, so-called transcription factors, which initiate this process either by directly binding to the DNA or by indirectly triggering the transcriptional activation. In the first case the proteins are typically classified based on their DNA-binding domain. Common domains are leucine zipper (LZ), the helix-turn-helix (HTH) motif, the basic-helix-loop-helix (HLH) motif, zinc-coordinating binding domains and β-scaffolds for minor groove contacts. The HLH superfamily of transcription factors contains alone more then 240 members that are further divided into different classes depending on the presence of other domains in addition to the HLH one (Figure 1.1) [1].

All HLH proteins except those of class V bear an additional basic region adjacent to the HLH motif, which enables the formed homo- or heterodimer to bind to specific hexameric DNA consensus sequences like the Ephrussi box (E-box) CANNTG and the N-box CACNAG, or to the Ets site GGAA/T, and thus to activate transcription. The Id (inhibitor of differentiation) proteins Id1-4 belong to the class V and lack the additional basic region but, nevertheless, form heterodimers with proteins from classes I and II, thus preventing the DNA binding of the latter. Using this inhibition mechanism, the Id proteins act as regulators of gene expression. The involvement of the HLH proteins in a wide range of developmental processes like cell differentiation, lineage commitment, and sex determination render them an interesting subject of research.

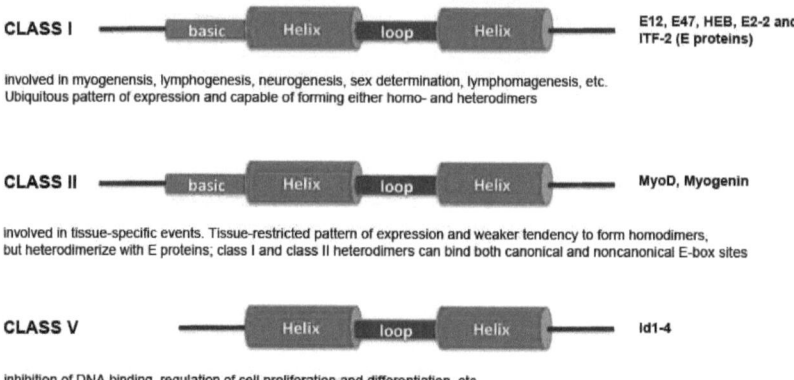

Figure 1.1: HLH proteins from classes I, II and V. Adapted from Deprez et al. [2] (basic: basic region).

The Id proteins lengths range from 119 amino acids for Id3 to 161 for Id4, which makes them rather small proteins. The Id HLH motif is highly conserved, especially in the region of helix-1 and helix-2, being the latter more conserved than helix-1 (Figure 1.2). Instead, the loop region is less conserved than the helices, with the exception of the N-terminal *PXXP* tetrapeptide that is common to all four Id proteins and might thus be important for the correct packing of helix-1 and helix-2 to build the HLH fold, which is essential for Id protein activity [3].

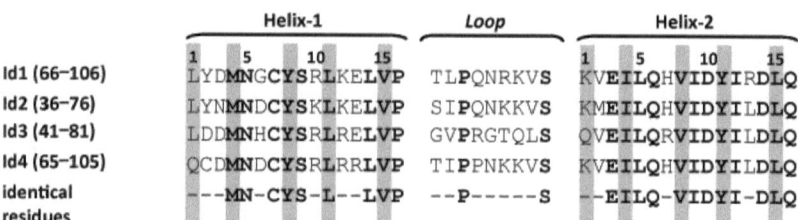

Figure 1.2: Amino acid sequences of the human Id HLH motifs. Adapted from Kiewitz et al. [4].

10

1.2 Mode of action and regulation of the Id proteins

The biological function of the Id proteins is based on their capacity to form heterodimers with bHLH transcription factors, in particular with the ubiquitous E proteins of the class I and with the tissue-specific factors of the class II. Due to the fact that the Id proteins do not posses the DNA-binding region, these heterodimers show no affinity for the DNA, which results in the lack of activation of DNA transcription (Figure 1.3). Therefore, the most important activity of the Id proteins in the cells is that of inhibiting the DNA binding, which leads to the inhibition of cell differentiation while inducing cell proliferation [5].

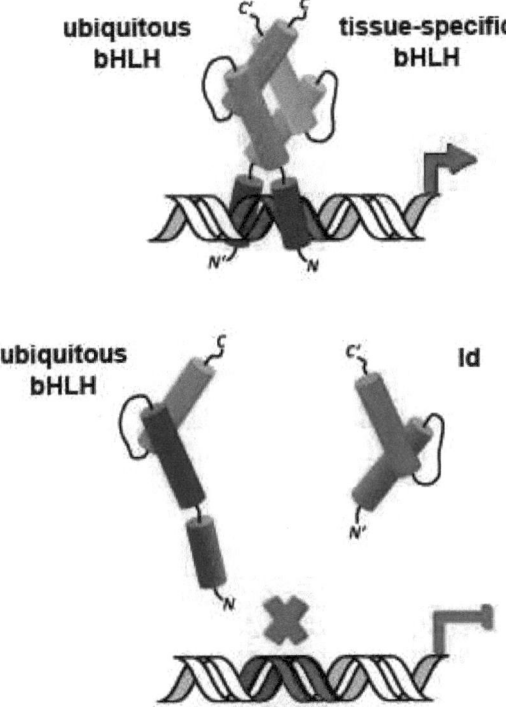

Figure 1.3: Inhibition of the activity of bHLH transcription factors by the Id proteins. Adapted from Kiewitz et al. [6].

Id3 protein activity is regulated at different levels, including gene and protein expression. Id3 protein phosphorylation and ubiquitination, alternative longer spliced variants and expression of Id3 binding patterns also play a role: for example, a modulation of Id3 activity is provided by a switch from the expression of Id3 to that of the longer spliced variant Id3L, as the latter cannot bind to E proteins [7–10]. Moreover, overexpressed E47 can sequester Id3 and thus inhibits its function. Phosphorylation of Id3 also affects its role in the cell cycle progression and proliferation [8].

The Id proteins are potential targets in cancer therapy [11], as they are deeply involved in pathways regulating proliferation, differentiation, angiogenesis, migration, invasion and cell-cell interaction [2, 12–14]. They are mainly overexpressed during embryogenesis and development as well as in tumor cells, whereas their expression significantly decreases in normal mature adult tissues [14, 15].

Id1 and Id3 are important for tumor angiogenesis, and their loss causes vascular defects in both xenografts and the more physiologically relevant tumor models. The angiogenic defect in Id-deficient mice may be attributed to combined downregulation of MMP-2 and fibroblast growth factor receptor 1 and $\alpha 6 \beta 4$ integrin. A partial decrease in Id levels significantly inhibits tumor invasion and metastasis because of reduced expression of MMPs and vascularization with consequent reduction of oxygen and nutrients supply to the tumor cells [16].

1.3 Structural studies of the Id HLH dimerization domain

The characterization and structural studies of the Id HLH domain is complicated by the presence of the Id HLH domain as dimer, tetramer or even higher-order oligomers in solution as evidenced by NMR measurements. One possible solution might be the formation of covalently bonded homodimers. Disulfide and lactam bridges represent possible covalent linkers.

The dissertation's work of Kiewitz has investigated the conformational features of the Id1 HLH fold [17]. For this purpose an interesting approach has been used which is based on the formation of the covalently linked Id1 HLH dimers by cysteine oxidation of analogs that bear GGC (Gly-Gly-Cys) moieties attached to either the N-terminal or the C-terminal ends of the HLH domain, respectively. The results of the oxidation experiments have shown that in the HLH dimer the parallel orientation of the first and second helices should be favored. Indeed thiol-disulfide exchange experiments did not produce the heterodimer, in which the helix-1 and the helix-2 of the corresponding monomer would adopt an anti-parallel conformation. This observation confirms a preferred parallel orientation of the helices in the Id1 HLH dimer, as it was also shown in the crystal structures of E47 and MyoD.

In another approach Id1 HLH analogs displaying modified loop regions and/or the retro sequence of helix-1 have been designed, in order to study the effect of the modulation of helix orientation on the HLH fold. For example, it was observed that the Id1 HLH domain with retro helix-1 presents a tighter helix packing than the native HLH. This leads to increased helical content revealing the importance not only of the interhelical side-chain packing, but also of the antiparallel helix dipole arrangement for the HLH fold stabilization. One reason for the natural parallel orientation of the helices in the HLH domain might be the control of folding stability: indeed, too high stability would be unfavorable for the occurrence of protein-protein interactions, which are necessary for the biological activity of the Id and other HLH proteins.

By using tyrosine/phenylalanine replacement Id1 HLH analogs showed a higher helical content than the native sequence [17]. This result can be justified by the fact that the natural tyrosine side chains are predominantly participating in hydrophobic interactions, which produce the formation and stabilization of the hydrophobic core of the protein fold [18]. However, the choice of Nature for tyrosine rather than for phenylalanine might have been dictated by the necessity of controlling the stability of the folding, in order to balance folding and unfolding states of the HLH domain.

The NMR spectroscopy is a powerful and useful technique to carry out structural studies of the protein domains. The HLH domain of Id2 was investigated by NMR spectroscopy and the NMR data have revealed the presence of a poorly defined N-terminal helix (helix-1) and of a stable C-terminal α-helix (helix-2) connected to each other by a short flexible region (loop). These data are in accordance with other reports upon the superior helix propensity of the second helix of the Id HLH domain compared to the first helix, as investigated by circular dichroism (CD) spectroscopy [19], another powerful tool to carry out structural and folding studies. In conclusion it was suggested that structural flexibility of the first helix constitutes an important structural feature occurring in the entire Id HLH protein family.

Furtheremore, in the dissertation's work of Kiewitz the ability of a ditopic receptor to bind to the Id HLH domain was investigated by fluorescence spectroscopy. The receptor consists of a fluorescent crown ether unit linked to a Cu^{II}-IDA (iminodiacetic acid) complex by an alkyl chain [17, 20]. In particular, the fluorescence response is triggered only upon a simultaneous binding of both moieties by a Lewis acid and a Lewis base group. The binding experiments have shown that the receptor binds to peptides representing the four Id HLH domains with $K_{0.5}$ values from 1.5 μM to 2.5 μM. In addition, the tetrapeptide $CxxR/K$ was identified as strong binding motif located in the first helix of the Id HLH domains. The receptor has also shown moderate selectivity towards the Id HLH motifs in comparison to the ones of the related HLH proteins MyoD and Max, which showed $K_{0.5}$ values of 6.0 μM and 5.2 μM, respectively.

All together, these results have proved that thiol oxidation and thiol-disulfide exchange assays are powerful tools to study the self-association of the Id HLH domain towards the formation of homodimers as well as the topology of the helices in the associated state. Moreover, modifications of the native Id sequences and the following CD and NMR spectroscopy studies are useful to determine the role of selected amino acids in the folding process. In addition, by using a luminescent artificial receptor, a unique tetrapeptide motif within the helix-1 could be identified, which might represent a starting point for the development of Id specific markers and small molecules capable to interfere with the Id proteins in cellular pathways.

1.4 Synthesis and conformational analysis of Id2 protein fragments

The Id2 protein is considered to be a promising target in tumor therapy, in particular for neuroblastoma, where Id2 interacts with the retinoblastoma protein (pRb) and inhibits its tumor suppression function.

In the dissertation work of Colombo the synthesis and conformational characterization of polypeptides derived from point mutations and N-/C-terminal truncations of Id2 have been discussed [21]. These peptides are useful tools to study the influence of different parts of the sequence on the structural properties of the protein. The synthesis of the series of Id2 analogs with variable length was performed by stepwise SPPS (solid phase peptide synthesis) using *Fmoc* chemistry. The synthesis of the C-terminal domain resulted to be difficult and dependent on the truncation point. Indeed, the C-terminal positions 124 and 110, but not 119, allowed the synthesis of large fragments. Moreover, Id2 fragments containing part of the C-terminus presented a general poor solubility under physiological conditions and a propensity to self-aggregate [19].

Further, by using amino-acid replacement, the residues Met-39/-62 and Cys-42 were found to play a role in the HLH fold, as their substitution with norleucine and serine, respectively, reduced the helix character of the HLH fold (residues 36-76). Moreover, it was found that the presence of the domain immediately following the C-terminal helix-2 was affecting the conformational properties of the HLH fold.

The role of the N- and C-terminal regions in all four Id proteins is not yet completely understood. It is known that phosphorylation of Ser-5 modulates the inhibitory activity of Id2 [22] and Id3 [5], and that the N-terminus of Id2 is necessary for ubiquitination [23]. Moreover, the HLH region of Id2 is required for the protein nuclear localization, whereas the sequence 103-119 in the central part of the C-terminus is necessary for the localization of the protein in the cytoplasm [24]. Therefore, the transport of Id2 from the nucleus to the cytoplasm is allowed by this C-terminal nuclear export signal (NES, residues 103-119). Based on the observation that, in general, peptides reproducing the C-terminal domain of the Id2 protein, which include the NES region, tend to aggregate, Colombo et al. [25] investigated the conformational properties of the NES region in the isolated form by using the synthetic Id2 fragment 103-124. Surprisingly, it was found that this fragment was insoluble in water and poorly soluble in water/alcohol mixtures. Transmission-electron microscopy (TEM) studies showed that the insoluble aggregates of this fragment were amorphous, despite the fact that the CD spectra recorded in water/alcohol mixtures revealed the presence of β-sheets. With the aim to increase the peptide solubility, a positive tag consisting of three lysine residues was added to the N-terminus or to the C-terminus of the Id2 fragment 103-124. As expected, the peptide solubility increased, but the effect of the N-tag was different from that of the C-tag: indeed, the N-tag only slightly increased the solubility, which, however, was sufficient to slow down the aggregation and allow to form ordered β-sheet fibrils capable to bind thioflavin T. In contrast, the C-tag attributed higher solubility to the Id2 fragment that was also found to adopt a helical conformation instead of a β-sheet conformation. Nevertheless, both Lys-tagged NES peptides could self-associate in a time-dependent manner and form ordered aggregates characterized by a fibrillar structure.

How a protein correlates to a specific biological or pathological event is one of the major challenges in medical research. Many cellular pathways are regulated by protein-protein interactions, thus the aberrant or inappropriate formation of protein complexes can lead to pathological events. It is yet unknown, however, whether the aggregating behavior of the C-terminal domain of Id2 might lead to pathologies other than cancer in Id2-protein overexpressing phenotypes.

1.5 Peptidomimetics as potential modulators of protein-protein interactions

The use of peptidomimetics as potential modulators of protein-protein interactions can allow to understand the mechanism of protein complex formation and to design new efficient therapeutically useful drugs. This approach has been studied in the dissertation's work of Colombo by using the unnatural amino acid 3-carboxy-cyclopentylglycine (**Cpg**). This scaffold displays two carboxylic groups separated by four carbon atoms, three of them are constrained in a cyclopentyl ring (Figure 1.4). The **Cpg** has been used as N,N-linkage of two Id helix-2 fragments to form covalent homodimers, and the effect of both enantiomers of Cpg on the conformation has been investigated by CD spectroscopy. The results showed that the Cpg-linked dimer of the Id1 helix-2 fragment 91-101 could interact with the native Id1 HLH motif and, depending on the Cpg configuration, lead to helix stabilization or destabilization of the latter.

In summary, the synthetic and conformational studies of Id2 protein fragments have revealed new interesting structural properties of the Id2 protein, in particular of the C-terminal domain and its effect on the adjacent HLH region. In addition, it could be shown, that the use of artificial building blocks for assembling short native Id protein fragments is suitable to develop Id-protein interacting molecules with modulatory effects.

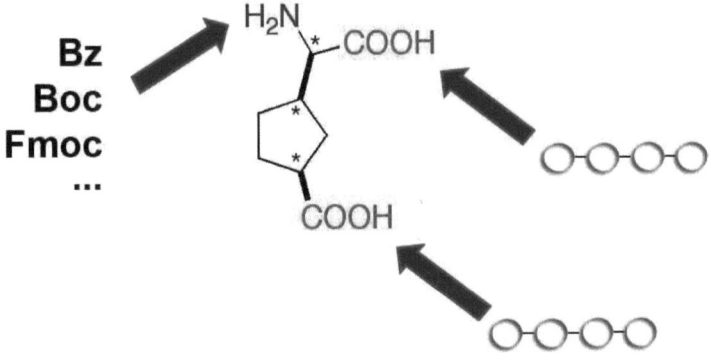

Figure 1.4: Cpg scaffold: the carboxylic groups are available for coupling to the N-terminus of peptides, while the amino group can be protected with different protecting groups and later coupled to peptides or other chemical moieties, including fluorescent probes.

1.6 Current structural information of Id3

The human Id3 protein displays the following 119-long amino acid sequence:
N-terminus domain: M^{1}KALSPVRGCYEAVCCLSERSLAIARGRGKGPAAEEPLSL^{40}
HLH domain: LDDMNHCYSRLRELVP^{56}GVPRGTQLS^{65} QVEILQRVIDYILDLQ^{81}
C-terminus domain: VVLAEPAPGPPDGPHLPIQTAELAPELVISMDKRSFCH^{119}

The HLH region spans residues 41-81, with helix-1 41-56, the loop 57-65 and helix-2 66-81. Cabrele and co-workers have reported the chemical synthesis of the full-length Id3 protein by Fmoc-solid-phase-methodology and its characterization by circular dichroism (CD) spectroscopy [27]. The synthetic Id3 protein has been found be able to form disulfide-bonded homodimers: indeed, five cysteine residues are present within the Id3 sequence. Based on the CD spectrum, about 30% of the residues are in helical conformation, most of which will be located in the HLH region. In contrast, the N- and C-terminal domains are likely to maintain significant degree of flexibility, but nevertheless, contribute to the overall folding, presumably by forming large loops to allow for long-range contacts between the three subdomains (N-terminus, HLH and C-terminus).

The helical character of the Id3 HLH structural domain has been also found to be sensitive to amino acid replacement within the two helices. In particular, whereas phenylalanine substitution of the helix-1 tyrosine residue at position 48 was fully tolerated, substitution of helix-2 tyrosine at position 76 partially destabilized the helical fold, suggesting that the phenol group of Tyr-76 is involved in favorable hydrogen bonds. In addition, whereas substitution of the helix-1 cysteine residue at position 47 with glycine was moderately tolerated in the isolated HLH peptide 41-81, it was deleterious in the larger Id3 peptide 1-81 that contains both the HLH region and the N-terminus [26]. This suggests that the N-terminus influences the conformation of the HLH domain. Therefore, it is important to understand the role played by the different subdomains of the Id3 protein in the overall folding and self-/hetero-association. To this purpose, in the Cabrele research group a series of fluorescent Id3 polypeptides fragments have been previously synthesized, in order to carry out FRET studies. The FRET pair of choice was Trp/Dns (dansyl): the Dns group was usually introduced at the N-terminus of the polypeptides, whereas Trp was introduced in place of a hydrophobic residue within the HLH motif or the C-terminus. The fluorescent peptides investigated so far in the Cabrele group are summarized in Figure 1.5.

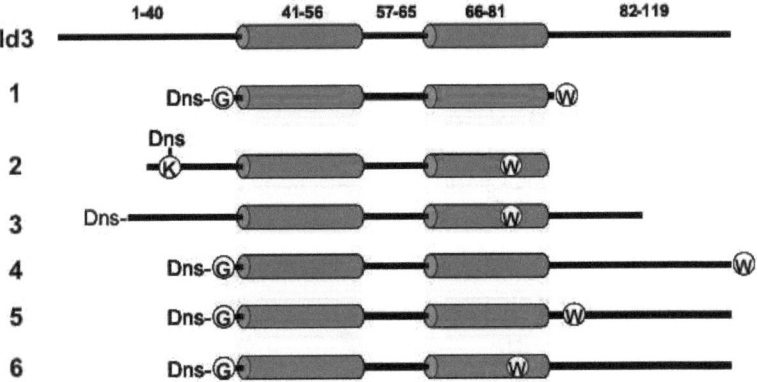

Figure 1.5: Fluorescent peptides investigated so far in the Cabrele group [23].

Whereas most of the labeling positions were sufficiently tolerated, leading to moderate conformational changes based on the CD spectra of the modified peptides, the introduction of Trp and of the Dns group at the C-end and N-end of the HLH fragment, respectively, was not tolerated, as this double-labeled peptide **1** was not able to fold into a helix-rich structure. Also the introduction of Trp as the additional C-terminal amidated residue was found to be deleterious, as analog 4 showed high content of unordered structure. This prevented its use for FRET studies. However, the FRET experiment performed provides some insight about the reciprocal spatial distribution of the HLH region and the N-terminal region: for example, the fact that peptides 2 and 3 were FRET active suggests that the N-terminal part of the polypeptide chain is likely to fold back toward the HLH region. Unfortunately, there was no information about the C-terminus, due tot the lack of suitable fluorescent Id3 polypeptide analogs.

The PhD work of Svobodova [27] has presented how the SPPS approach using Fmoc chemistry can be successfully applied for the synthesis of the full-length Id3 protein as well as truncated and modified analogues. The conformational studies by circular dichroism (CD) on all synthesized Id3 polypeptides revealed a complex interplay between the Id3 subdomains. Furthermore, the FRET studies on the fluorescent-labeled Id3 fragments suggest that the N-terminus folds back to the compact HLH motif. These results contributed to planning the investigations of this dissertation work in order to provide further knowledge.

1.7 Scope and objectives

Starting from the results previously collected in the Cabrele's group [17, 26, 27] on the synthesis and conformation of Id3 and related analogs, the presented PhD thesis is dedicated to the better understanding of the folding properties of the HLH dimerization domain of the Id3 protein. In particular, it will be focused on three different approaches to study and control the self-association of the Id3 HLH domain towards homodimers formation. Moreover, studies of the behavior of the HLH dimerization domain of the Id3 and Id4 proteins upon environmental changes will be presented and discussed.

One of the three approaches is based on the use of thiol-containing analogs and oxidation/disulfide reshuffling assays, as discussed before [17]. New analogs of the Id3 HLH peptide 41-81 were designed to perform oxidation studies towards the formation of well-structured covalent dimers and the results will be discussed in the next chapters.

Another approach to control the self-association of the Id3 HLH domain towards homodimers, which will be presented in this thesis, is based on electrostatic interactions between a positively charged tag of an Id3 HLH analog and the negatively charged one of another analog.

The last proposed approach is based on the use of fluorescent peptide analogs to be investigated by FRET spectroscopy. For this purpose based on results previously collected [26] new fluorescent Id3 polypeptides analogs were designed and the results will be presented in the next chapters.

The Id3 protein fragments that were required for the studies were synthesized by solid-phase methodology using the *Fmoc*-chemistry. Purification of the crude products was carried out by preparative HPLC. Identification and characterization of the pure peptides were based on MALDI-TOF-MS and analytical HPLC techniques. Structural investigation was carried out by using CD and fluorescence spectroscopy.

References

[1] Atchley, W. R.; Fitch, W. M., A natural classification of the basic helix-loop-helix class of transcription factors. *Proc Natl Acad Sci U S A* **1997**, 94 (10), 5172-6.

[2] Deprez, P. Y.; Sumida, T.; Coppe, J. P., Helix-loop-helix proteins in mammary gland development and breast cancer. *J Mammary Gland Biol Neoplasia* **2003**, 8 (2), 225-39.

[3] Pesce, S.; Benezra, R., The loop region of the helix-loop-helix protein Id1 is critical for its dominant negative activity. *Mol Cell Biol* **1993**, 13 (12), 7874-80.

[4] Kiewitz, S. D.; Cabrele, C., Synthesis and conformational properties of protein fragments based on the Id family of DNA-binding and cell-differentiation inhibitors. *Biopolymers* **2005**, 80 (6), 762-74.

[5] O'Toole, P. J.; Inoue, T.; Emerson, L.; Morrison, I. E.; Mackie, A. R.; Cherry, R. J.; Norton, J. D., Id proteins negatively regulate basic helix-loop-helix transcription factor function by disrupting subnuclear compartmentalization. *J Biol Chem* **2003**, 278 (46), 45770-6.

[6] Kiewitz, S. D.; Kruppa, M.; Riechers, A.; Konig, B.; Cabrele, C., Recognition of the helix-loop-helix domain of the Id proteins by an artificial luminescent metal complex receptor. *J Mol Recognit* **2008**, 21 (2), 79-88.

[7] Deed, R. W.; Hara, E.; Atherton, G. T.; Peters, G.; Norton, J. D., Regulation of Id3 cell cycle function by Cdk-2-dependent phosphorylation. *Mol Cell Biol* **1997**, 17 (12), 6815-21.

[8] Forrest, S. T.; Taylor, A. M.; Sarembock, I. J.; Perlegas, D.; McNamara, C. A., Phosphorylation regulates Id3 function in vascular smooth muscle cells. *Circ Res* **2004**, 95 (6), 557-9.

[9] Berse, M.; Bounpheng, M.; Huang, X.; Christy, B.; Pollmann, C.; Dubiel, W., Ubiquitin-dependent degradation of Id1 and Id3 is mediated by the COP9 signalosome. *J Mol Biol* **2004**, 343 (2), 361-70.

[10] Deed, R. W.; Armitage, S.; Norton, J. D., Nuclear localization and regulation of Id protein through an E protein-mediated chaperone mechanism. *J Biol Chem* **1996**, 271 (39), 23603-6.

[11] Fong, S.; Debs, R. J.; Desprez, P. Y., Id genes and proteins as promising targets in cancer therapy. *Trends Mol Med* **2004**, 10 (8), 387-92.

[12] Sikder, H. A.; Devlin, M. K.; Dunlap, S.; Ryu, B.; Alani, R. M., Id proteins in cell growth and tumorigenesis. *Cancer Cell* **2003**, 3 (6), 525-30.

[13] Norton, J. D., ID helix-loop-helix proteins in cell growth, differentiation and tumorigenesis. *J Cell Sci* **2000**, 113 (Pt 22), 3897-905.

[14] Coppe, J. P.; Smith, A. P.; Desprez, P. Y., Id proteins in epithelial cells. *Exp Cell Res* **2003**, 285 (1), 131-45.

[15] Lyden, D.; Young, A. Z.; Zagzag, D.; Yan, W.; Gerald, W.; O'Reilly, R.; Bader, B. L.; Hynes, R. O.; Zhuang, Y.; Manova, K.; Benezra, R., Id1 and Id3 are required for neurogenesis, angiogenesis and vascularization of tumor xenografts. *Nature* **1999**, 401 (6754), 670-7.

[16] Kim, D. S.; Franklyn, J. A.; Boelaert, K.; Eggo, M. C.; Watkinson, J. C.; McCabe, C. J., Pituitary tumor transforming gene (PTTG) stimulates thyroid cell proliferation via a vascular endothelial growth factor/kinase insert domain receptor/inhibitor of DNA binding-3 autocrine pathway. *J Clin Endocrinol Metab* **2006**, 91 (11), 4603-11.

[17] Kiewitz S. D., PhD Dissertation: Structural investigations of the Id helix-loop-helix dimerization domain. University of Regensburg, **2007**.

[18] Harbury, P. B.; Zhang, T.; Kim, P. S.; Alber, T., A switch between two-, three-, and four-stranded coiled coils in GCN4 leucine zipper mutants. *Science* **1993**, 262 (5138), 1401-7.

[19] Colombo, N.; Cabrele, C., Synthesis and conformational analysis of Id2 protein fragments: impact of chain length and point mutations on the structural HLH motif. *J Pept Sci* **2006**, 12 (8), 550-8.

[20] Kiewitz, S.D.; Kruppa, M.; Riechers, A.; König, B.; Cabrele, C., Recognition of the Helix-Loop-Helix Domain of the Id Proteins by an Artificial Luminescent Metal Complex Receptor. *J. Mol. Recognit.* **2008**, 21, 79-88.

[21] Colombo, N., PhD Dissertation: Synthesis and conformational analysis of polypeptides related to the inhibitor of the DNA binding and cell differentiation Id2. University of Regensburg, **2006**.

[22] Hara, E.; Hall, M.; Peters, G., Cdk2-dependent phosphorylation of Id2 modulates activity of E2A-related transcription factors. *EMBO J* **1997**, 16 (2), 332-42.

[23] Fajerman, I.; Schwartz, A. L.; Ciechanover, A., Degradation of the Id2 developmental regulator: targeting via N-terminal ubiquitination. *Biochem Biophys Res Commun* **2004**, 314 (2), 505-12.

[24] Kurooka, H.; Yokota, Y., Nucleo-cytoplasmic shuttling of Id2, a negative regulator of basic helix-loop-helix transcription factors. *J Biol Chem* **2005**, 280 (6), 4313-20.

[25] Colombo, N.; Schroeder, J.; Cabrele, C., A Short Id2 Protein Fragment Containing the Nuclear Export Signal Forms Amyloid-Like Fibrils. *Biochem. Biophys. Res. Commun.* **2006**, 346, 182-187.

[26] Svobodova, J., PhD Dissertation: Id3, Inhibitor of DNA Binding and Cell Differentiation: Synthesis and Conformational Analysis of the Full-Length Protein and its Truncated Analogues. University of Regensburg, **2007**.

[27] J. Svobodova, and C. Cabrele, Stepwise Solid-Phase Synthesis and Spontaneous Homodimerization of the Helix-Loop-Helix Protein Id3. *Chem Bio Chem* **2006**, 7, 1164-68.

Chapter 2

Design and synthesis of Id3 protein fragments for conformational studies

2.1 Introduction

The chemical approach for the synthesis of peptides and small proteins is an attractive alternative to the expression in bacteria or mammalian cells, as it allows to carry out chemical manipulations of the native target by the use of backbone surrogates, unnatural amino acids, conformational constraints or site-specific labels. All of these synthetic tools are helpful for the analysis of the peptide/protein structure and folding. This chapter describes the synthesis of some fragments of Id3 protein, used for the conformational and self-association studies of the HLH domain presented in this thesis, and shows the difficulties related to the synthesis of these polypeptides.

2.2 Solid-phase synthesis of the Id3 protein fragments

Stepwise solid-phase methodology is widely applied for the synthesis of peptides with small to medium chain length. The accumulation of byproducts related to the lack of quantitative coupling/deprotection reactions represents the major problem [1]. Nevertheless, since the introduction of the solid-phase peptide syntheses (SPPS) by Merrifield [2] in 1963, a number of stepwise synthesis of polypeptides and small proteins both with the Boc (tert-butoxycarbonyl) and the *Fmoc* (9-fluorenylmethoxycarbonyl) strategies have been reported [3–12]. Thus, we decided to investigate the latter approach applied to the Id3 protein. The polypeptide chain was automatically synthesized on Rink Amide MBHA resin, whose loading was 0.45 mmol/g, or Rink amide Chem-Matrix®, whose loading was 0.54 mmol/g. Double couplings of N^{α}-*Fmoc*-protected amino acids were accomplished with *in situ* activation by HBTU (*O*-(1-benzotriazolyl)-1,1,3,3-tetramethyluronium hexafluorophosphate), HOBt and DIPEA in DMF/NMP (N,N-dimethylformamide/1-methyl-2-pyrrolidinone); the *Fmoc* group was removed by piperidine/DMF/NMP. To control the quality of the growing chain, small amounts of peptide were detached from the resin. The cleavage mixture consisted of trifluoroacetic acid (TFA) containing 10% (v/v) scavengers: water, triisopropylsilane (TIS) and ethanedithiol (EDT). The peptide fragments of Id3 were then purified and characterized by mass spectrometry and analytical HPLC (Table 2.1 and Figure 2.1).

Table 2.1: Analytical data of the synthesized Id3 polypeptides **II.1-6** (**II.3a** crude).

	Synthetic Id3 polypeptides			
Product	Sequence	$MW_{calc.}$ (Da)	MW_{found} (Da)	t_r (min.)
II.1	Fmoc-[K-84]-(38-86)-NH$_2$	5858.9	–	–
II.2	Ac-[G-47, K-84]-(47-99)-NH$_2$	5839.8	5840.6	22.7^a
II.3a	Fmoc-[K-84]-(45-92)-NH$_2$	5587.5	5585.0	21.2^b
II.3b	Fmoc-[G-37, K-84]-(37-92)-NH$_2$	6432.5	6432.5	28.1^a
II.4	Fmoc-[K-47]-(20-92)-NH$_2$	8203.6	–	–
II.5	Ac-(48-92)-NH$_2$	5037.9	5039.6	24.7^b
II.6	Ac-(31-63)-NH$_2$	3663.2	3663.9	16.4^a
II.7	(41-81)-NH$_2$	4796.6	4794.0	30.0^c
II.8	Ac-CGG-[S^{47}]-(41-81)-GGG-NH$_2$	5211.0	5211.0	23.7^d
II.9	Ac-[S^{47}]-(41-81)-GGC-NH$_2$	5039.8	5038.7	24.7^d
II.10	Ac-R$_5$-(41-81)-NH$_2$	5619.6	5618.8	22.1^e
II.11	Ac-E$_5$-(41-81)-NH$_2$	5484.2	5485.8	23.5^e

^{a}Gradient: 20% ACN for 5 min., 20-80% ACN over 40 min.
^{b}Gradient: 40% ACN for 5 min., 40-85% ACN over 40 min.
^{c}Gradient: 20% ACN for 5 min., 20-70% ACN over 35 min.
^{d}Gradient: 30% ACN for 5 min., 30-70% ACN over 30 min.
^{e}Gradient: 30% ACN for 5 min., 30-80% ACN over 35 min.

Peptides **II.1-4** represent intermediates of the synthesis of fluorescence labeled Id3 polypeptides reproducing the regions 1-86, 38-92, 1-92 and 1-99. The introduction of the fluorescent probe (FAM or TAMRA) was planned either near the C-terminus (position 84) or near the N-terminus (position 37 and 47) of the HLH domain. For this purpose Lys(Mtt) was used, as the Mtt protecting group can be orthogonally removed from the side chain, thus allowing the selective reaction with the fluorophore. The very low homogeneity of the peptides **II.1-4** (Figure 2.1) prevented any attempt to continue the sequence till the planned position and to perform the desired labeling. In particular, in the case of the peptides **II.1** and **II.4** the homogeneity was so bad that it was not even possible to purify and characterize them. Only the peptides **II.2** and **II.3b** could be purified and characterized by MALDI-TOF-MS and HPLC. However, the crude products of the corresponding fluorescently labeled peptides were not clean enough and it was not possible to obtain pure fluorescent products.

In contrast, in the case of peptides **II.5-7** we obtained good homogeneity (Figure 2.1), which allowed labeling and purifying easily the crude products. Based on these results, it seems that the synthetic problems were dependent on the starting position. In fact, the reproduction of the entire HLH domain, residues 41-81, seems to be difficult by starting from positions in the C-terminal domain (86, 92, and 99) compared to position 81, which corresponds to the end of helix-2. Obviously, also the length of the polypeptide chain plays a role: for example, the shorter fragment 45-92 (**II.3a**) was obtained as the major product. Instead, after elongation the fragment 37-92 (**II.3b**) was obtained with lower homogeneity (Figure 2.2). This suggests that the synthetic problems encountered in the preparation of the fragment 37-92 occurred during elongation from residue 44 to residue 37. The very low homogeneity obtained for the fragment 20-92 (peptide **II.4**) (Figure 2.1) confirms the suggestion. It seems that the final part of helix-1 and the beginning of N-terminus present a difficult sequence to synthesize probably due to the aggregation of the peptide chains on the resin. We performed longer coupling reactions with HOBt/DIC as coupling reagents and the addition of the base DIPEA 20 minutes after the addition of the coupling reagents. However this procedure did not lead to a significant improvement of the quality of the crude product.

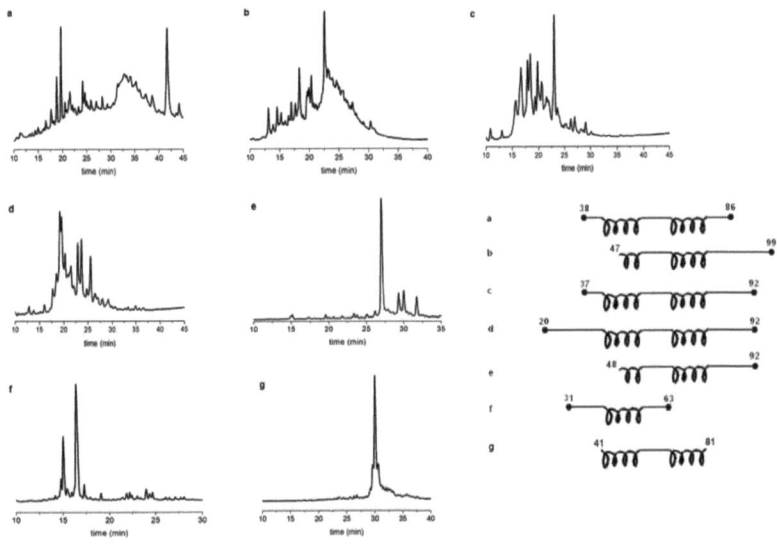

Figure 2.1: HPLC profiles of the crude peptides **II.1, II.2, II.3b, II.4, II.5, II.6** and **II.7**: (a-g) Fmoc-[K-84]-(38-86)-NH$_2$, Ac-[G-47, K-84]-(47-99)-NH$_2$, Fmoc-[G-37, K-84]-(37-92)-NH$_2$, Fmoc-[K-47]-(20-92)-NH$_2$, Ac-(48-92)-NH$_2$, Ac-(31-63)-NH$_2$, H-(41-81)-NH$_2$.

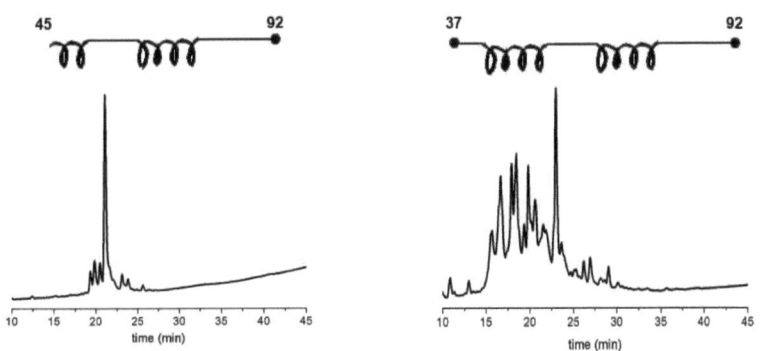

Figure 2.2: HPLC profiles of the crude peptides **II.3a-b**: Fmoc-[K-84]-(45-86)-NH$_2$ (*left*), Fmoc-[G-37, K-84]-(37-92)-NH$_2$ (*right*).

The peptide **II.6** was synthesized with the Rink amide ChemMatrix® to test the quality of this resin compared to the polystyrene-based one. The ChemMatrix® resin in particular presents better swelling property with DMF and also TFA, and this could improve the efficiency of the couplings and the cleavages. However, this resin did not lead to a significant improvement of the quality of the crude product. It is important to underline the presence of two major peaks in the HPLC profile of peptide **II.6** (Figure 2.1 f). The first peak around 15 minutes corresponds to the Met-oxidized species (+ 16 Da) and this is understandable because the HPLC profile was recorded before the reductive treatment by TMSBr/EDT/TFA solution. Indeed this treatment was used to avoid any loss of product because of oxidation.

The HLH motif 41-81 of Id3 (peptide **II.7**) was obtained with good homogeneity supporting the results already obtained in previous works [13]. This peptide was further elongated to obtain the R5- and E5-tagged analogs (**II.10** and **II.11**). One noteworthy byproduct during the synthesis of the HLH motif was its Met-oxidized species (+ 16 Da), which was recovered by reduction with TMSBr/EDT/TFA as we did before in the case of peptide **II.6**.

The peptides **II.8** and **II.9** were synthesized for studies of structural preference of the Id3 HLH dimer, which will be described in the next chapter. In the case of peptide **II.9** that bears the cysteine at the C-terminus, the homogeneity results lower (Figure 2.3) than that of the analog **II.8** that bears the cysteine at the N-terminus.

Both peptides were treated with TMSBr/EDT/TFA after the synthesis to reduce the Met-oxidized species.

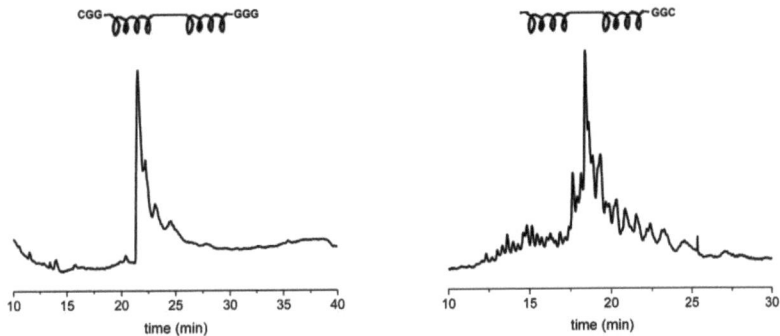

Figure 2.3: HPLC profiles of the crude peptides **II.8** Ac-CGG-(41-81)-GGG-NH$_2$ (*left*) and **II.9** Ac-(41-81)-GGC-NH$_2$ (*right*).

The peptides **II.10** and **II.11** were synthesized to study and control the self-association of the HLH domain by electrostatic interactions, which will be discuss later. The HLH domain 41-81 was assembled in the automatic synthesizer and the last five amino acids were coupled manually. The crude products are showed in Figure 2.4. The HPLC profile of the crude peptide **II.11** shows a splitted major peak, as a deleted product is present, where one Glu is missed as confirmed by MALDI analysis.

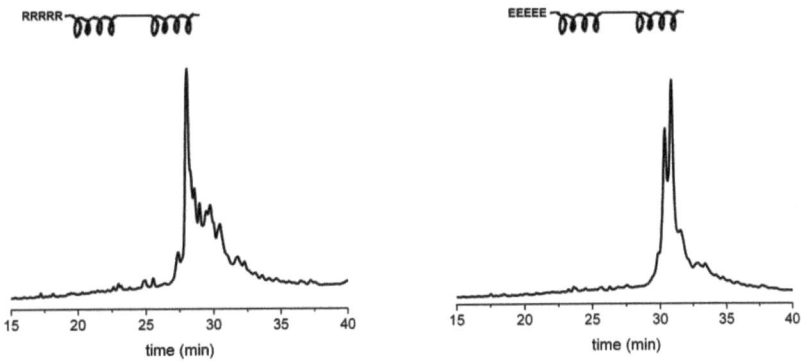

Figure 2.4: HPLC profiles of the crude peptides **II.10** Ac-R$_5$-(41-81)-NH$_2$ (*left*) and **II.11** Ac-E$_5$(41-81)-NH$_2$ (*right*).

2.3 Conclusions

Synthetic peptides derived from amino acid replacement and different chain length of Id3 are useful tools to study the structural preference and the self-association of the Id3 HLH domain. In this chapter, we have reported on the synthesis of a series of different Id3 analogs used for the conformational and self-association studies in this thesis, which were prepared by stepwise SPPS using *Fmoc* chemistry.

The position 86, 92 and 99 in the C-terminal domain were found to be critical for assembling sequences containing the complete HLH domain 41-81. In particular, the synthesis of the N-terminal part of helix-1 and the beginning of the N-terminal domain resulted to be difficult, as confirmed by the low homogeneity of the crude products obtained. This is probably related to the aggregation of the peptide chains on the resin.

During the synthesis of the HLH motif we have observed the formation of its Met-oxidized species as a byproduct. The reductive treatment with TMSBr/TFA/EDT has demonstrated to be a good procedure to recover the fully reduced species and avoid loss of product. The synthesis of the loop-helix-1 and the beginning of the N-terminus resulted to be easy and proceeded without difficulties. Indeed, the crude product resulted quite clean and easy to purify.

The introduction of a cysteine residue at the beginning of the peptide synthesis has been accompanied by lower homogeneity of the crude compared to the introduction of cysteine at the N-terminus of the sequence. This may be due to the formation of side products related to oxidation events during the synthesis as well as to the acidic sensitivity of the GC peptide bond under acidic conditions [14].

References

[1] Kent, S. B., Chemical synthesis of peptides and proteins. *Annu Rev Biochem* **1988**, 57, 957-89.

[2] Merrifield; R. B., Solid Phase Peptide Synthesis. I. The Synthesis of a Tetrapeptide. *J Am Chem Soc* **1963**, 85, 2149-2154.

[3] Clark-Lewis, I.; Moser, B.; Walz, A.; Baggiolini, M.; Scott, G. J.; Aebersold, R., Chemical synthesis, purification, and characterization of two inflammatory proteins, neutrophil activating peptide 1 (interleukin-8) and neutrophil activating peptide. *Biochemistry* **1991**, 30 (12), 3128-35.

[4] Goetz, M.; Geoffre, S.; Busetta, B.; Manigand, C.; Nespoulous, C.; Londos-Gagliardi, D.; Guillemain, B.; Hospital, M., Synthesis and CD studies of an 88-residue peptide containing the main receptor binding site of HTLV-I SU-glycoprotein. *J Pept Sci* **1997**, 3 (5), 347-53.

[5] Goud, N. A.; McKee, R. L.; Sardana, M. K.; DeHaven, P. A.; Huelar, E.; Syed, M. M.; Goud, R. A.; Gibbons, S. W.; Fisher, J. E.; Levy, J. J.; et al., Solid-phase synthesis and biologic activity of human parathyroid hormone (1-84). *J Bone Miner Res* **1991**, 6 (8), 781-9.

[6] Gutte, B.; Merrifield, R. B., The synthesis of ribonuclease A. *J Biol Chem* **1971**, 246 (6), 1922-41.

[7] Abiko, T.; Sekino, H., Synthesis of prothymosin alpha deduced from nucleotide sequence of the murine cDNA and its effect on the impaired T lymphocytes of uremic patients. *Biotechnol Ther* **1993**, 4 (3-4), 213-20.

[8] Bonetto, V.; Massignan, T.; Chiesa, R.; Morbin, M.; Mazzoleni, G.; Diomede, L.; Angeretti, N.; Colombo, L.; Forloni, G.; Tagliavini, F.; Salmona, M., Synthetic miniprion PrP106. *J Biol Chem* **2002**, 277 (35), 31327-34.

[9] Ferrer, M.; Woodward, C.; Barany, G., Solid-phase synthesis of bovine pancreatic trypsin inhibitor (BPTI) and two analogues. A chemical approach for evaluating the role of disulfide bridges in protein folding and stability. *Int J Pept Protein Res* **1992**, 40 (3-4), 194-207.

[10] Fukuda, H.; Irie, K.; Nakahara, A.; Ohigashi, H.; Wender, P. A., Solid-phase synthesis, mass spectrometric analysis of the zinc-folding, and phorbol ester-binding studies of the 116-mer peptide containing the tandem cysteine-rich C1 domains of protein kinase C gamma. *Bioorg Med Chem* **1999**, 7 (6), 1213-21.

[11] Kaiser, T.; Luppa, P.; Voelter, W., Solid-phase synthesis, isolation and analysis of a mouse protein, the macrophage migration inhibitory factor. *J Chromatogr A* **1999**, 852 (1), 189-95.

[12] Dong, C. Z.; Romieu, A.; Mounier, C. M.; Heymans, F.; Roques, B. P.; Godfroid, J. J., Total direct chemical synthesis and biological activities of human group IIA secretory phospholipase A2. *Biochem J* **2002**, 365 (Pt 2), 505-11.

[13] Svobodova, J. Id3, Inhibitor of DNA Binding and Cell Differentiation: Synthesis and Conformational Analysis of the Full-Length Protein and its Truncated Analogues. University of Regensburg, **2007**.

[14] Kang, J.; Richardson, J. P.; Macmillan, D., 3-Mercaptopropionic acid-mediated synthesis of peptide and protein thioesters. *Chem Commun (Camb)* **2009**, (4), 407-9.

Chapter 3

Covalent dimers of the Id3 HLH domain

3.1 Introduction

The HLH domain of the Id proteins is prone to the formation of dimers and tetramers, thus leading to a non-homogeneous solution, although the equilibrium is more shifted towards the tetrameric form at peptide concentrations above 10 μM [1]. The simultaneous presence of different non covalent oligomers renders conformational studies, in particular based on NMR spectroscopy, very difficult. This has been shown in the case of the bHLH domain of MyoD, which was studied by NMR spectroscopy as disulfide-bonded homodimer [2]. The disulfide bond was obtained by oxidation of the native cysteine residue located near the C-end of the helix-1 of the MyoD HLH domain. However, the NMR structure of the oxidized form of MyoD bHLH resulted to be different from that found in the crystal structure of the non-covalent bHLH homodimer complexed to the DNA [3]. Indeed, the latter displays the parallel four-helix bundle that is characteristic of the DNA-binding bHLH transcription factors [4, 5]. Instead, the disulfide-bonded MyoD bHLH dimer adopted an antiparallel orientation of the intrachain helices, which led to an antiparallel four-helix bundle. The reason for this different helix packing might be the presence of the non-native disulfide bond, or the absence of DNA, which might allow the formation of the antiparallel bundle. In any case, this supports that the helices of the HLH domain may arrange in two different ways.

The Id proteins do not bind the DNA, as they lack the DNA-binding region. Therefore, the dimerization of the Id HLH domain should be also able to adopt both a parallel or antiparallel topology. The preferred topology for the Id HLH domain was unclear, until the crystal structure of the Id2 HLH domain [6] and the NMR solution structure of the Id3 HLH domain [7] were recently published. Both structures show a parallel four-helix bundle. Before these structures were available, we were studying the structural preference of the Id1 [8] and Id3 HLH domain by using different approaches. As for the MyoD bHLH domain, we tried to stabilize the dimeric form of the Id HLH domain over higher-order oligomers by forming a disulfide-bonded dimer. The results are shown here for the Id3 HLH domain.

3.2 Peptide design

The Id3 HLH domain contains one native cysteine at position 47, which is located in the middle of the helix-1. However, we decided not to use this residue to form an intermolecular disulfide bond, to avoid possible constraints in the helix packing. Therefore, we introduced a cysteine residue beyond the HLH domain, separated from the N- or C-terminus by the insertion of a short diglycine spacer. This strategy has been used by other groups for the determination of the topology of helix bundles [9]. As shown in Table 3.1, peptide **III.1** carries a CGG moiety at the N-terminus of the HLH domain, while peptide **III.2** carries this moiety at the C-terminus. In addition, peptide **III.1** contains a GGG moiety at the C-terminus: the reason for this is the necessity of having different masses for the two cysteine-containing analogs of the Id3 HLH domain, in order to facilitate the identification of potential **III.1/III.2** heterodimers by mass spectrometry. Furtheremore, in all peptides the native cysteine at position 47 was substituted by serine to prevent the formation of undesired oxidation products.

Table 3.1: Amino acid sequences of the synthetic Id3 HLH analogs used for dimerization and thiol-disulfide exchange studies.

No.	Description	Sequence
III.1	Ac-CGG-[S^{47}]-(41-81)-GGG Id3	Ac-**CGG**-LDDMNHSYSRLRELV-PGVPRGTQLS-QVEILQRVIDYILDLQ-**GGG**-NH_2
III.2	Ac-[S^{47}]-(41-81)-GGC Id3	Ac-LDDMNHSYSRLRELV-PGVPRGTQLS-QVEILQRVIDYILDLQ-**GGC**-NH_2

At the end of the synthesis, the N-terminus was acetylated and the peptides were cleaved from the resin, purified by semipreparative RP-HPLC, and characterized by analytical RP-HPLC and MALDI-TOF mass spectrometry (Figure 3.1 and Table 3.2). After the TFA cleavage, both peptides were additionally treated with the mixture TMSBr/EDT/TFA to reduce the oxidized methionine residues.

Table 3.2: Analytical data of the synthetic Id3 HLH analogs **III.1** and **III.2**.

Product	Sequence	$MW_{calc.}$ (Da)	MW_{found} (Da)	t_r (min.)^a
III.1	Ac-CGG-[S^{47}]-(41-81)-GGG Id3	5211.0	5211.0	23.7
III.2	Ac-[S^{47}]-(41-81)-GGC Id3	5039.8	5038.7	24.7

^{a}Gradient: 30% ACN for 5 min., 30-70% ACN over 30 min.

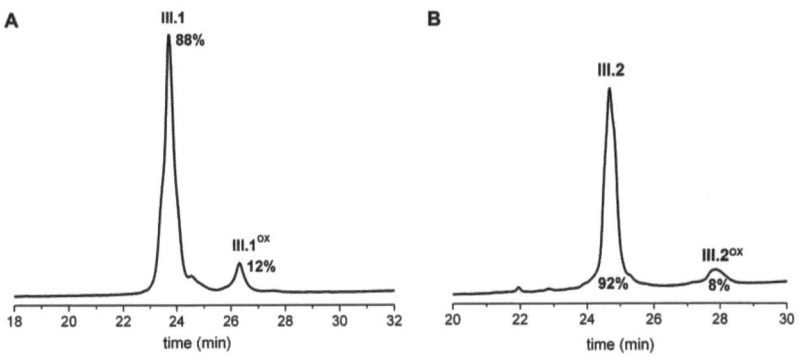

Figure 3.1: HPLC profile of purified peptides **III.1** (**A**) and **III.2** (**B**).

Figure 3.1 shows the HPLC profiles of the purified peptides **III.1** and **III.2**. In both cases, a small amount of the oxidized species is already present. In the case of peptide **III.1** 12% of monomer is oxidized, and in the case of peptide **III.2** 8%. It seems that the oxidation is a favored process in both peptides.

Oxidation of each of these peptides should generate covalently linked homodimers, in which the helices adjacent to the disulfide bridge adopt a parallel orientation. If such oxidation would run smoothly, it would be an indication, that the two N-terminal helices or the two C-terminal helices tend to interact with each other in a parallel topology. In contrast, if they would prefer an antiparallel topology, the formation of the covalent homodimer would be unfavorable, whereas a covalent heterodimer would be favored. To prove these hypotheses, we first performed oxidation experiments towards the building of homodimers (**III.1**^{OX} and **III.2**^{OX}). Successively, we carried out two types of experiments towards the formation of heterodimers: the first experiment was based on a disulfide reshuffling in a redox buffer (Figure 3.2). The second experiment was based on a thiol-disulfide exchange under inert atmosphere (Figure 3.3).

Different propensity to form homo- and heterodimers will give an indication of the preferred helix orientation in the dimerization process of the Id3 HLH domain.

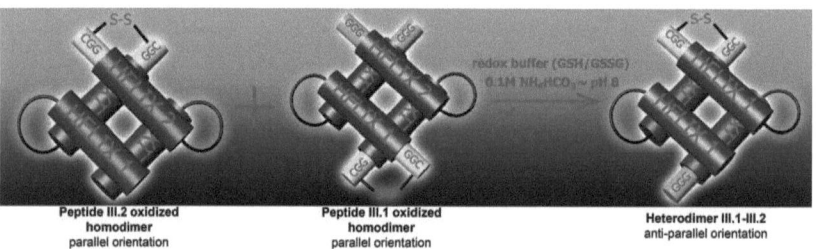

Figure 3.2: Disulfide reshuffling mediated by GSH/GSSG redox buffer. *GSH reduced glutathione; GSSG oxidized glutathione.*

Figure 3.3: Thiol/disulfide exchange under inert atmosphere.

The disulfide reshuffling assay was accomplished in a solution containing a redox buffer (based on the oxidized and reduced species of glutathione) in a slightly basic buffered solution (0.1 M ammonium carbonate at pH ∼ 8).

To perform the thiol/disulfide exchange we mixed one oxidized peptide (**III.2**^{OX}) with the monomer of the other (**III.1**) under inert atmosphere.

3.3 Results

3.3.1 Formation of the oxidized homodimers of Id3 HLH analogs

The synthetically prepared peptides **III.1** and **III.2** were dissolved in ammonium carbonate (pH ∼ 8) at the concentration of 0.7 mg in 500 µL and subjected to air oxidation, in order to produce the corresponding covalently linked homodimers **III.1**^{OX} and **III.2**^{OX}. The disulfide bond formation was monitored by HPLC over 36 h. However, already after 20 h the HPLC peak of reduced monomer **III.2** almost disappeared and a new peak corresponding to the oxidized peptide appeared at a higher retention time (Figure 3.4). The monomer **III.1** did not completely oxidize, as confirmed by the presence of a peak at 23.7 min, probably because there is a thiol/disulfide equilibrium or the disulfide building is less favored than for **III.2**. Therefore, it was necessary to purify the oxidized peptide **III.1**^{OX} to remove the not oxidized species (around 36%).

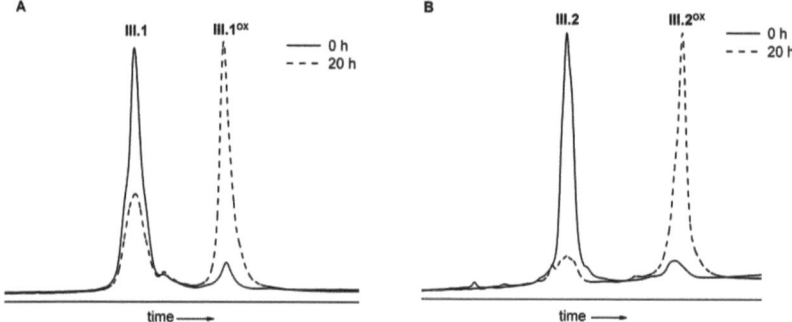

Figure 3.4: Monitoring of the air oxidation of peptides **III.1** (A) and **III.2** (B) by HPLC. The solid line represents the HPLC profile of an aliquot taken immediately after dissolving the peptide in ammonium carbonate buffer. The dash line is the HPLC profile after 20 h oxidation upon air oxygen.

After oxidation, and purification of **III.1^{OX}**, the peptides were lyophilized and the presence of the oxidized species was additionally confirmed by mass spectrometry (Table 3.3).

Table 3.3: Mass characterization of the lyophilized homodimers **III.1^{OX}** and **III.2^{OX}**.

	$MW_{mon,calc}$ (Da)	$MW_{mon,found}$ (Da)	$MW_{dimer,calc}$ (Da)	$MW_{dimer,found}$ (Da)
III.1	5211.0	5211.0	10420.0	10420.5
III.2	5039.8	5038.7	10077.6	10078.0

3.3.2 Disulfide reshuffling and thiol-disulfide exchange assays

To further investigate the preference for one distinct orientation, the two homodimers of peptides **III.1** and **III.2** were mixed under conditions that allow reduction of existing disulfide bonds and formation of new ones. The reshuffling mediated by glutathione buffer was monitored by HPLC and MS. After 30 min from the beginning of the assay the mass measurements showed the reduced species of **III.1** and **III.2**, and also their glutathione adducts, as indicated by the masses of the monomers plus 307 Da. The corresponding HPLC profile showed the two peaks of the homodimers (26.2 min for **III.1^{OX}** and 27.7 min for **III.2^{OX}**), preceded by the two peaks of the monomers (23.7

min for **III.1** and 24.3 min for **III.2**). In addition, there are two peaks at lower retention times, which are attributed to the glutathione adducts (Figure 3.5). The HPLC profiles and the mass measurements did not show the heterodimers formation between **III.1** and **III.2**.

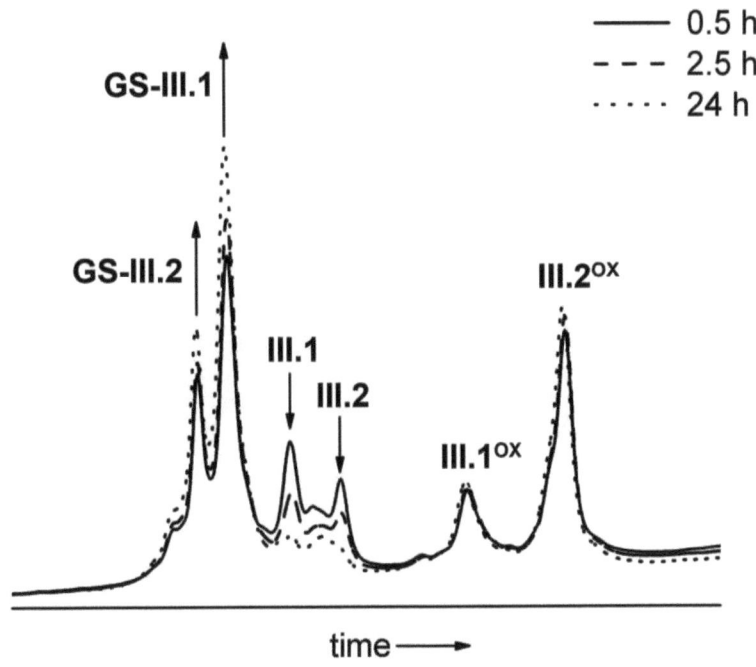

Figure 3.5: HPLC runs of aliquots of the disulfide reshuffling mixture (**III.1**^{OX}/**III.2**^{OX} 1:1) taken at 0.5 h, 2.5 h and 24 h.

After 2 h 30 min and 24 h the HPLC profile did not change dramatically, but it seems that the homodimer of **III.2** is more stable than the homodimer of **III.1**. After 24 h the peak of monomer **III.1** almost disappeared, and it formed the glutathione adduct, as confirmed by MALDI-TOF-MS.

In the thiol-disulfide exchange assay, the HPLC profile after 30 min showed again the two peaks of the homodimers (**III.1**^{OX} and **III.2**^{OX}) preceded by the peak of monomer **III.1** and two small peaks close to it (Figure 3.6). The HPLC profiles after 2 h 30 min and 24 h are practically unchanged (data not shown). Also in this case the heterodimers formation between **III.1** and **III.2** was not detected, as indicated by the HPLC profile and the mass measurements.

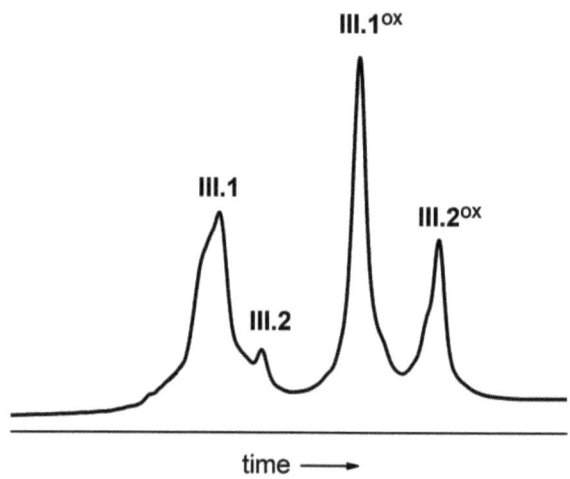

Figure 3.6: HPLC runs of aliquots of the thiol-disulfide exchange mixture (**III.1**/**III.2**^{OX} about 2:1) taken at 0.5 h.

3.4 Discussion

The two possible folding models for the covalent homodimer of peptide **III.1** are shown in Figure 3.7. In one model the four helices are all parallel, with the two helix-1 fragments being N-terminally linked, whereas in the other model the two helix-1 fragments N-terminally linked are parallel to each other but anti-parallel to the two helix-2 fragments. Thus, the latter model resembles the one of the disulfide-bridged homodimer of MyoD [2]. The same models are possible for the homodimer of **III.2**, with the difference that the two helix-2 fragments are C-terminally linked. The fact that the air oxidation of both peptides, **III.1** and **III.2**, proceeded easily and almost to completeness, suggests that the formation of the corresponding homodimers was favorable in both cases, and that the parallel orientation of the two helix-1 segments in **III.1**^{OX} or the two helix-2 segments in **III.2**^{OX} is the preferred one.

Figure 3.7: Schematic illustration of the possible folding models for the oxidized form of peptide **III.1** (*left*: all-parallel conformation, *right*: 1-1' parallel conformation 1-2 anti-parallel conformation). Adapted from Kiewitz [8].

In contrast, the generation of the mixed disulfide-bonded dimer between **III.1** and **III.2** would propose two different models of helix packing, which are schematically drawn in Figure 3.8.

Figure 3.8: Schematic illustration of the possible folding models for the disulfide-bonded mixed dimer between **III.1** and **III.2** (*left*: 1-2' anti-parallel conformation 1,2 parallel conformation, *right*: 1-2' anti-parallel conformation 1-2 anti-parallel conformation). Adapted from Kiewitz [8].

In both possible dimeric packing models helix-1 of **III.1** and helix-2 of **III.2** would adopt an anti-parallel orientation to each other. The fact that the disulfide reshuffling and thiol-disulfide exchange assays did not show the formation of covalent heterodimers suggests that an interchain anti-parallel arrangement of helix-1 and helix-2 is unfavored. Moreover, the observation that the homodimer of **III.1** was converted for the major part into mixed disulfides with the glutathione compared to the homodimer of **III.2**, suggests that the covalently linked helix-2 segments are more stable than the covalently linked helix-1 segments. This, in turn, may indicate a tighter self-packing of the helix-2 fragments compared to a moderate or weak self-packing of the helix-1 segments.

3.5 Conclusions

We have prepared two disulfide-bonded Id3 HLH homodimers that build an all-parallel four-helix bundle, in agreement with the NMR study of the Id3 HLH region [7] (Figure 3.9). Despite the fact that antiparallel four-helix bundles are more common in nature than parallel ones, the HLH region of the Id proteins represents one of the few examples of parallel helix topology, and it can be thus used as model or scaffold for the de-novo design and mimicry of such protein fold.

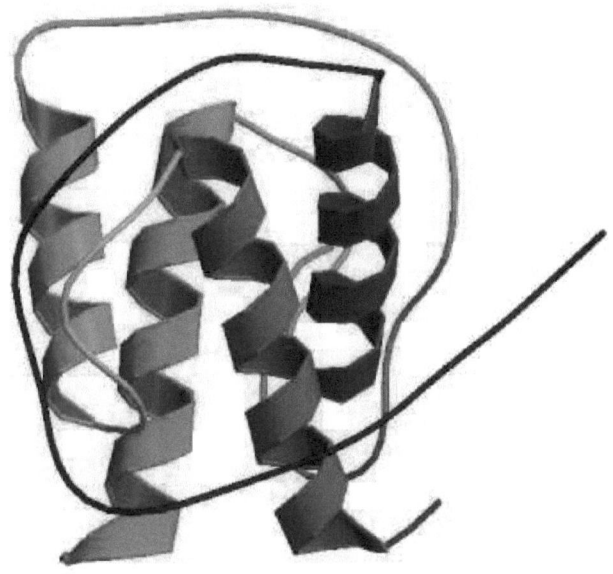

Figure 3.9: Solution NMR structure of the helix-loop-helix domain of human Id3 protein, Northeast Structural Genomics Consortium Target HR3111A (PDB ID: 2LFH) [7].

References

[1] Fairman, R.; Beran-Steed, R. K.; Anthony-Cahill, S. J.; Lear, J. D.; Stafford, W. F., 3rd; DeGrado, W. F.; Benfield, P. A.; Brenner, S. L., Multiple oligomeric states regulate the DNA binding of helix-loop-helix peptides. *Proc Natl Acad Sci U S A* **1993**, 90 (22), 10429-33.

[2] Starovasnik, M. A.; Blackwell, T. K.; Laue, T. M.; Weintraub, H.; Klevit, R. E., Folding topology of the disulfide-bonded dimeric DNA-binding domain of the myogenic determination factor MyoD. *Biochemistry* **1992**, 31, 9891-903.

[3] Ma, P. C.; Rould, M. A.; Weintraub, H.; Pabo, C. O., Crystal structure of MyoD bHLH domain-DNA complex: perspectives on DNA recognition and implications for transcriptional activation. *Cell* **1994**, 77 (3), 451-9.

[4] Ellenberger, T.; Fass, D.; Arnaud, M.; Harrison, S. C., Crystal structure of transcription factor E47: E-box recognition by a basic region helix-loop-helix dimer. *Genes Dev* **1994**, 8 (8), 970-80.

[5] Ferre-D'Amare, A. R.; Prendergast, G. C.; Ziff, E. B.; Burley, S. K., Recognition by Max of its cognate DNA through a dimeric b/HLH/Z domain. *Nature* **1993**, 363 (6424), 38-45.

[6] Wong, M. V.; Jiang, S.; Palasingam, P.; Kolatkar, P. R., A divalent ion is crucial in the structure and dominant-negative function of ID proteins, a class of helix-loop-helix transcription regulators. *PLoS One* **2012**, 7 (10), e48591.

[7] Eletsky, A., Wang, D., Kohan, E., Janjua, H., Acton, T.B., Xiao, R., Everett, J.K., Montelione, G.T., Szyperski, T.. Solution NMR Structure of the Helix-loop-Helix

Domain of Human ID3 Protein, Northeast Structural Genomics Consortium Target HR3111A. **DOI**:10.2210/pdb2lfh/pdb

[8] Kiewitz, S. D., PhD Dissertation: Structural investigations of the Id helix-loop-helix dimerization domain. Universitt Regensburg, Regensburg, **2007**.

[9] Lai, J. R.; Fisk, J. D.; Weisblum, B.; Gellman, S. H., Hydrophobic core repacking in a coiled-coil dimer via phage display: insights into plasticity and specificity at a protein-protein interface. *J Am Chem Soc* **2004**, 126 (34), 10514-5.

Chapter 4

Studies of the self-association of oligo-Arg/Glu-tagged analogs of the Id3 HLH domain

4.1 Introduction

The Id proteins differ from their bHLH dimerization partners like MyoD and E47 in the basic region N-terminal to the HLH domain, which is required to bind specific DNA motifs and to stabilize the formed tertiary complex between DNA and the bHLH dimer. Crystal structures of such ternary complexes are known for MyoD [1], E47 [2] and other bHLH factors. From these structures it is known, that the additional DNA-binding region adopts a helical conformation and strongly interacts with the DNA helix. The Id proteins lack this region and are thus unable to interact with the DNA. Nevertheless, they are able to form stable homodimers, as shown for Id2 [3] and Id3 [4, 5]. This suggests, that the self-association of the HLH domain, which is highly conserved in the HLH superfamily [6], does not require the presence of the DNA-binding region and the DNA-complex formation. This is also supported by the NMR structure of the MyoD homodimer [7] and by the X-ray structure of a chimera E47-Max homodimer [8], both in the absence of DNA. Apparently, the two positively-charged N-terminal DNA-binding regions do not prevent self-association, suggesting that there are no unfavorable charge repulsions.

In analogy to the bHLH domain, we have investigated the effect of a charged subdomain at the N-terminus: for this purpose, we have elongated the N-terminus of the Id3 HLH domain by a penta-arginine. In addition, we have prepared an analog containing a N-terminal penta-glutamate and investigated the self- and hetero-association upon the control of attracting or repulsive electrostatic interactions.

The Id3 bHLH analogs containing the penta-arginine (basic region), **IV.1**, and the penta-glutamate (acidic region), **IV.2**, are shown in Figure 4.1.

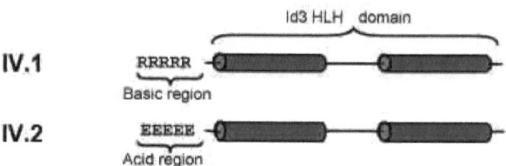

Figure 4.1: Cartoons of the Id3 HLH analogs containing a basic and acidic region at the N-terminus.

4.2 Peptide synthesis

The two analogs of the Id3 HLH motif were synthesized by SPPS. The amino acid sequences of the peptides **IV.1** and **IV.2** are reported in the Table 4.1.

Table 4.1: Amino acids sequences of the peptides **IV.1** and **IV.2**.

No.	Description	Sequence
IV.1	Ac-R$_5$-(41-81) Id3	Ac-**RRRRR**-LDDMNHCYSRLRELV-PGVPRGTQLS-QVEILQRVIDYILDLQ-NH$_2$
IV.2	Ac-E$_5$-(41-81) Id3	Ac-**EEEEE**-LDDMNHCYSRLRELV-PGVPRGTQLS-QVEILQRVIDYILDLQ-NH$_2$

Once the synthesis was completed, the N-terminus was acetylated and the peptides were cleaved from the resin, purified by semipreparative RP-HPLC, and characterized by analytical RP-HPLC and MALDI-TOF mass spectrometry (Table 4.2 and Figure 4.2).

After the TFA cleavage, both peptides had to be additionally treated with the mixture TMSBr/EDT/TFA in order to reduce the oxidized methionine residues.

Table 4.2: Analytical data of the pure synthetic Id3 HLH analogs **IV.1** and **IV.2**.

No.	Description	$MW_{calc.}$ (Da)	MW_{found} (Da)	t_r (min.)
IV.1	Ac-R$_5$-(41-81)-NH$_2$	5619.6	5618.8	22.1
IV.2	Ac-E$_5$-(41-81)-NH$_2$	5484.2	5485.8	23.5

Gradient: 30% ACN for 5 min., 30-80% ACN over 35 min.

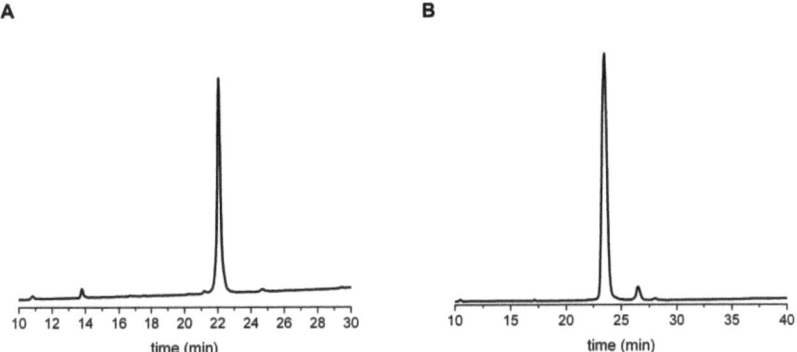

Figure 4.2: HPLC profile of purified peptides **IV.1** (**A**) and **IV.2** (**B**).

4.3 Results and discussion

4.3.1 Conformational studies by CD spectroscopy

The conformation of the Id3 HLH analogs was investigated by CD spectroscopy in phosphate buffer (100 mM, pH = 7.1) and compared with that of the native one (**II.7**). The peptide concentration was 42 μM. We measured also the conformation of the mixtures of the peptides **IV.1** and **IV.2** in different ratios (Table 4.3), where the total concentration of the mixtures was 25 μM.

Table 4.3: Mixtures of the peptides **IV.1** and **IV.2** at different ratios.

Mixture	% peptide IV.1	% peptide IV.2
9:1 (**IV.1/IV.2**)	90	10
7:3 (**IV.1/IV.2**)	70	30
1:1 (**IV.1/IV.2**)	50	50
3:7 (**IV.1/IV.2**)	30	70
1:9 (**IV.1/IV.2**)	10	90

The peptide **IV.1** gave a α-helix-like CD spectrum with much lower intensity than that of the native sequence of the Id3 HLH (Figure 4.3). Therefore five Arg residues at the N-terminus have a negative effect on the helical structure that decreased from 46% to 13% with an increase in the strand fraction from 12% to 35% and disordered fraction from 25% to 32% (Table 4.4). In contrast, peptide **IV.2** gave α-helix-like CD spectrum with much higher intensity than that of the native Id3 HLH (Figure 4.3), which indicates a positive effect of the negatively charged peptides on the helical structure that increased from 46% to 76% at the expense of the disordered, strand and turn fractions.

Table 4.4: Secondary structure elements composition of the peptides **II.7**, **IV.1** and **IV.2** (the CD spectra deconvolution was performed by using the algorithm Contin [9] from the online server Dichroweb [10].

	T (°C)	Helix (%)	Strand (%)	Turn (%)	Disordered (%)
II.7	20	46	12	17	25
IV.1	20	13	35	20	32
IV.2	20	76	2	12	10

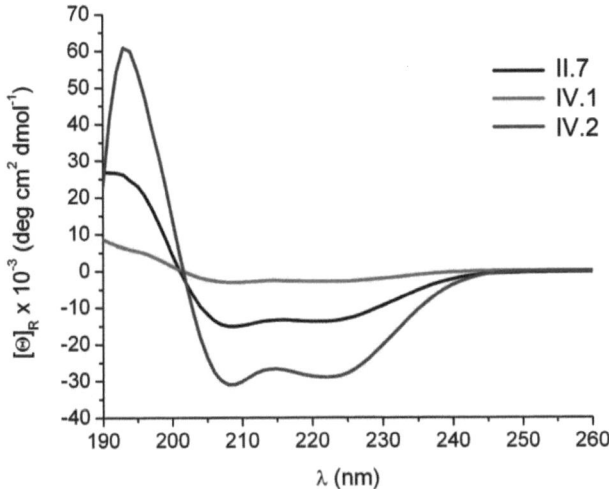

Figure 4.3: CD spectra of peptides **IV.1** and **IV.2** and of the native Id3 HLH (**II.7**) at 42 µM concentration in phosphate buffer (100 mM, pH 7.1).

The negative effect of the N-terminal positive charges (peptide **IV.1**) on the HLH fold is possibly due to a destabilization of the helix-1 macro-dipole, the steric hindrance of the Arg side chains and also to electrostatic repulsions between the positively charged side chains of Arg and positive side chains of other amino acids, for example those of the Arg side chain in the loop region. Indeed, in the crystal structure of the MyoD bHLH dimer bound to DNA it can be seen, that the loop is proximal to the basic region and that the Arg side chain of the MyoD loop interacts with the DNA [1]. On the other hand, the positive effect of the N-terminal negative charges (peptide **IV.2**) on the HLH fold is possibly due to the stabilization of the helix-1 macro-dipole, to the less steric hindrance of the Glu side chains in comparison to the Arg side chains, and to favorable electrostatic interactions between the negatively charged side chains of Glu and the positive charged ones of other amino acids.

When the two peptides were mixed together, the two negative bands at 208 nm and 222 nm become more intense upon the increasing of the concentration of peptide **IV.2**

(Figure 4.4A-B). The deconvolution of the CD spectra of Figure 4.4A gave the results reported in Table 4.5. These data suggest an increase in the helical content upon the increasing of the concentration of peptide **IV.2**. However, it is interesting to note that the helical content of peptide **IV.2** is lower (76%) than in the mixture 1:9 (**IV.1/IV.2**) (85%), probably because in the mixture there are favorable electrostatic interactions which further favor the helix packing. The comparison of the CD curves of the mixtures with molar fraction (X) of peptide **IV.2** of 0.3 and 0.5 with the CD curve of the native Id3 HLH shows that the intensity of the latter curve is in between the other two curves (Figure 4.4B). Moreover, the secondary structure elements composition of the mixture with $X = 0.3$ is close to the composition of the native Id3 HLH (Table 4.5).

Further, we expressed the interaction between the two mixed peptides in terms of the difference in the ellipticity at 222 nm of the experimental CD curve of each mixture compared to the theoretical CD curve, which arises from the arithmetic sum of the CD curves of the peptides alone. The difference $\Delta\Theta_{222nm}$ was calculated by using the following equation:

$$\Delta\Theta_{222nm} = |\Theta|_{222exp} - |\Theta|_{222theor} = |\Theta|_{222exp} - \{(|\Theta|_{222,X=1} \times X) + [|\Theta|_{222,X=0} \times (1-X)]\}$$

The difference ellipticity value was then plotted as a function of X (molar fraction of peptide **IV.2**) in order to obtain the Jobs plot (Figure 4.4D). The fitted curve shows two maxima (Figure 4.4D). This indicates the formation of two different oligomers with different stoichiometry: one is of 1:4 between peptide **IV.1** and **IV.2** and the other is of 4:1.

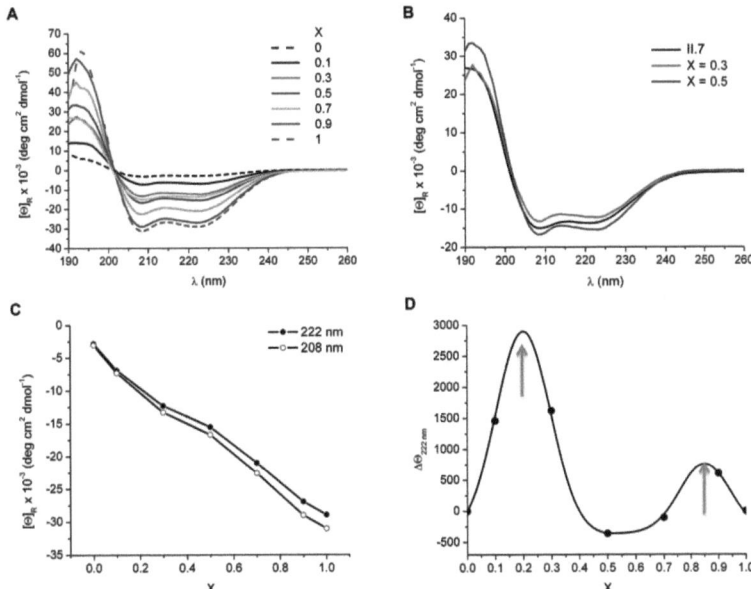

Figure 4.4: CD spectra of peptides **IV.1**, **IV.2** and their mixtures (**A**) as well as the mixtures with X = 0.3, X = 0.5 and the native Id3 HLH (X = molar fraction of peptide **IV.2**) (**B**) in phosphate buffer (100 mM, pH 7.1). The two minima at 208 nm and 222 nm as a function of the molar fraction of peptide **IV.2**, X (**C**). The Jobs plot of the difference ellipticity at 222 nm as a function of the molar fraction of peptide **IV.2**, X (**D**).

Table 4.5: Secondary structure elements composition of the peptides **IV.1**, **IV.2** and their mixtures (the CD spectra deconvolution was performed by using the algorithm Contin [9] from the online server Dichroweb [10]).

	T (°C)	Helix (%)	Strand (%)	Turn (%)	Disordered (%)
1:0 (**IV.1/IV.2**)	20	13	35	20	32
9:1 (**IV.1/IV.2**)	20	27	25	19	29
7:3 (**IV.1/IV.2**)	20	46	14	16	24
1:1 (**IV.1/IV.2**)	20	55	8	15	22
3:7 (**IV.1/IV.2**)	20	72	2	12	14
1:9 (**IV.1/IV.2**)	20	86	1	4	9
0:1 (**IV.1/IV.2**)	20	76	2	12	10

4.3.2 Thermal dissociation of the helix bundle

The thermal dissociation of the native Id3 HLH and their analogues in phosphate buffer (100 mM, pH = 7.1) was studied by CD spectroscopy by measuring the changing of the minima at 222 nm with the temperature from 20°C to 90°C. The concentration of the peptides was the same used before in the conformational studies. After the melting, we performed also a cooling to see if the denaturation is reversible. To draw the melting curve in the Figure 4.5A-B we plotted the value of % folded structure as function of the temperature. The % folded structure is a relative amount calculated by taking the molar ellipticity value at 222 nm at 20°C as the value of 100% folded structure.

The melting curve of the peptides and the mixtures had a sigmoid shape (Figure 4.5A-B). The inflection point of these curves represent the melting temperature of the folded structure. The **IV.1/IV.2** complexes are not easily thermally denatured, as they appear to be more than 50% folded at 90°C (Figure 4.5A). The melting temperature of the mixed oligomers increased upon increasing of the concentration of peptide **IV.2**, reaching a plateau around 70°C and 0.7 molar fraction of **IV.2**, as we can see from the Figure 4.5C.

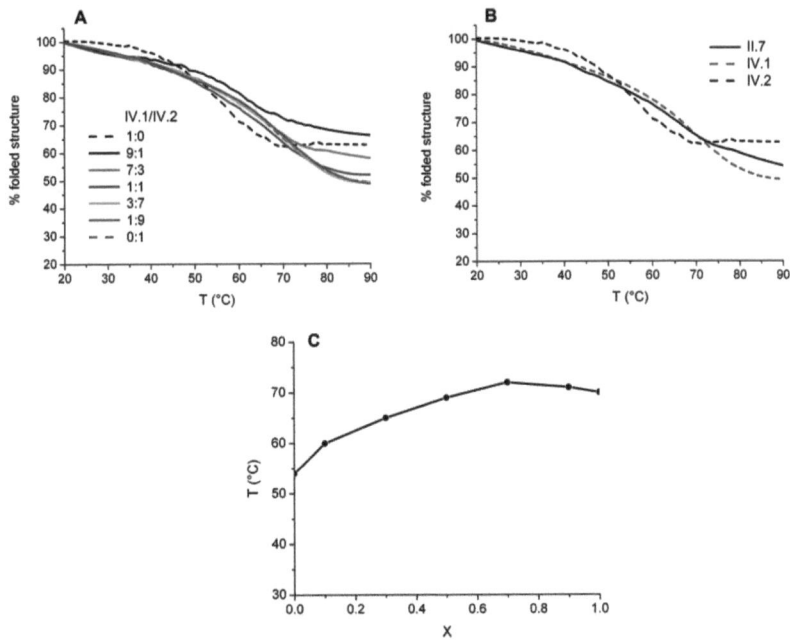

Figure 4.5: Melting curves of peptides **IV.1**, **IV.2** and their mixtures (**A**) and the peptides **IV.1**, **IV.2** and the native Id3 HLH (**B**) in phosphate buffer (100 mM, pH 7.1). The melting temperature of the mixtures as a function of the molar fraction of peptide **IV.2**, X (**C**).

The native fold was fully recovered after cooling in the case of peptide **IV.2** and all the mixtures except 9:1 (**IV.1/IV.2**), indicating that the thermal dissociation was reversible (data not shown). Interestingly, for peptides **IV.1** and the mixture 9:1 (**IV.1/IV.2**) the thermal dissociation is not completely reversible, showing a loss of 26% and 18%, respectively, of the folded structure after cooling.

The analog **IV.1** results thermodynamically less stable than the natural Id3 HLH, as indicated the decrease in melting temperature from 65°C of the native Id3 HLH to 54°C of the peptide **IV.1** (Table 4.6). In contrast, the analog **IV.2** presents higher melting temperature, 70°C, and this means that it is thermodynamically more stable than the natural Id3 HLH and the arginine-tagged analog (Table 4.6).

Table 4.6: Melting temperatures of peptides **IV.1** and **IV.2** and their mixture and of the natural Id3 HLH (**II.7**), calculated by the derivative of the sigmoid curves.

Mixture	T_m (°C)
1:0 (**IV.1/IV.2**)	54
9:1 (**IV.1/IV.2**)	60
7:3 (**IV.1/IV.2**)	65
1:1 (**IV.1/IV.2**)	69
3:7 (**IV.1/IV.2**)	72
1:9 (**IV.1/IV.2**)	71
0:1 (**IV.1/IV.2**)	70
II.7 (natural Id3 HLH)	65

4.4 Conclusions

The CD studies of the tagged analogs have demonstrated as the presence of five Arg residues at the N-terminus of the Id3 HLH motif has a negative effect, as it destabilized the folding and increased the content of unordered, strand and turn structures. In contrast, the presence of five Glu residues at the N-terminus has a positive effect, as it stabilized the folding and increased the content of the helical structure.

The CD results of the mixture of these two tagged analogs suggest an interaction between them. Indeed, the two analogs are apparently able to associate and form different oligomers, in particular pentamers like **IV.1/IV.2** 1:4 and **IV.1/IV.2** 4:1.

The thermal unfolding/refolding of the two analogs and their mixed oligomers confirms the positive effect of the N-terminal negative charges (5 Glu) and the negative effect of the N-terminal positive charges (5 Arg) on the HLH fold. In particular, the N-terminal negatively charged Id3 HLH results thermally more stable than the native Id3 HLH, as demonstrated by its higher melting temperature (T_m). In contrast, the lower melting temperature (T_m) of the N-terminal positively charged Id3 HLH compared to the native Id3 HLH suggests a lower thermal stability and, moreover, it is partially unable to refold after thermal unfolding.

Further studies are necessary to improve the knowledge on the association equilibrium. For example an idea could be the FRET studies of two analogs of **IV.1** and **IV.2** containing a donor and a quencher, respectively.

References

[1] Ma, P. C.; Rould, M. A.; Weintraub, H.; Pabo, C. O., Crystal structure of MyoD bHLH domain-DNA complex: perspectives on DNA recognition and implications for transcriptional activation. *Cell* **1994**, 77 (3), 451-9.

[2] Ellenberger, T.; Fass, D.; Arnaud, M.; Harrison, S. C., Crystal structure of transcription factor E47: E-box recognition by a basic region helix-loop-helix dimer. *Genes Dev* **1994**, 8 (8), 970-80.

[3] Liu, J.; Shi, W.; Warburton, D., A cysteine residue in the helix-loop-helix domain of Id2 is critical for homodimerization and function. *Biochem Biophys Res Commun* **2000**, 273 (3), 1042-7.

[4] Wibley, J.; Deed, R.; Jasiok, M.; Douglas, K.; Norton, J., A homology model of the Id-3 helix-loop-helix domain as a basis for structure-function predictions. *Biochim Biophys Acta* **1996**, 1294 (2), 138-46.

[5] Svobodova, J.; Cabrele, C., Stepwise solid-phase synthesis and spontaneous homodimerization of the helix-loop-helix protein Id3. *Chembiochem* **2006**, 7 (8), 1164-8.

[6] Chavali, G. B.; Vijayalakshmi, C.; Salunke, D. M., Analysis of sequence signature defining functional specificity and structural stability in helix-loop-helix proteins. *Proteins* **2001**, 42 (4), 471-80.

[7] Starovasnik, M. A.; Blackwell, T. K.; Laue, T. M.; Weintraub, H.; Klevit, R. E., Folding topology of the disulfide-bonded dimeric DNA-binding domain of the myogenic determination factor MyoD. *Biochemistry* **1992**, 31, 9891-903.

[8] Ahmadpour, F.; Ghirlando, R.; De Jong, A. T.; Gloyd, M.; Shin, J. A.; Guarne, A., Crystal structure of the minimalist Max-E47 protein chimera. *PLoS One* **2012**, 7 (2), e32136.

[9] Provencher, S. W.; Glockner, J., Estimation of globular protein secondary structure from circular dichroism. *Biochemistry* **1981**, 20 (1), 33-7.

[10] Whitmore, L.; Wallace, B. A., DICHROWEB, an online server for protein secondary structure analyses from circular dichroism spectroscopic data. *Nucleic Acids Res* **2004**, 32 (Web Server issue), W668-73.

Chapter 5

A FRET study on self-association of Id3 protein fragments

5.1 Introduction

In the previous chapters two different approaches has been presented to study and control the self-association of the Id3 HLH domain towards homodimers. In this chapter a third approach will be presented, which is based on the use of FRET spectroscopy. Id3 protein fragments labeled with carboxyfluorescein/carboxytetramethylrhodamine (FAM/TAMRA) donor/acceptor pair (Föster radius 55 Å) were prepared and studied by FRET. We used an incomplete HLH sequence lacking part of helix-1. This fragment is expected to be only partially folded, which should reflect the early folding states of the Id3 protein. Indeed, the structure of the monomeric form is assumed to be quite flexible and to contain only locally ordered motifs. Further, we decided to truncated the helix-1, as helix-2 displays higher intrinsic helix propensity and, thus, might represent a locally ordered element in the early folding states: for this reason, we additionally prepared two fluorescent analogs reproducing helix-1, to study possible interactions between helix-2 and the nascent helix-1. The effect of the fluorophores introduction on the conformation of the native sequence was also investigated by CD spectroscopy. Moreover, the self-recognition between partially and fully folded HLH domains was studied by fluorescence spectroscopy.

5.2 Design and solid-phase synthesis of the fluorescently labeled Id3 protein fragments

In previous studies Svobodova [1, 2] prepared large polypeptide fragments containing the FRET pair Trp/Dns (Dns = dansyl). In the present study, instead of using the FRET pair Trp/Dns, alternative donors and acceptors were tested, especially to avoid the use of the Dns group that was found to reduce the water solubility of the peptides and made the purification by preparative HPLC difficult. We chose FAM/TAMRA as FRET pairs because both FAM and TAMRA can be easily and selectively introduced in the peptide chain by acylation of a suitable amino group, either at the N-terminus or at the side chain of a lysine residue. In our case, the fluorophore was introduced at the position 84 in the C-terminal region, which is naturally occupied by leucine, through the side chain of a lysine residue. We decided also to try the introduction of the fluorophore at position 76 in the region of helix-2, which is naturally occupied by tyrosine, through the side chain of a lysine residue. Another position for labeling was chosen in the loop and precisely at position 61, which is naturally occupied by glycine.

We have decided to use an N-terminally truncated HLH domain, as a model for a partially folded HLH domain, with the aim to mimic the first events of folding and self-association or hetero-association. This is the reason why we have chosen a quite flexible fragment considering that it lacks a part of helix-1.

The Id3 peptides containing the donor/acceptor pair FAM/TAMRA used in this study are summarized in the Figure 5.1. For the same sequence two different analogs with labeling in the same position were synthesized, one containing the donor (FAM) and the other the acceptor (TAMRA). In addition, the unlabeled Id3 segment was prepared to examine the effect of the fluorophores on the conformation (Table 5.1). Moreover, the corresponding native segment was prepared to see the effect of the single-residue replacement with lysine. Further, the analogs (41-101)-Id3 (Table 5.1) was used for the self-recognition studies between partially and fully folded HLH domains by fluorescence spectroscopy.

Table 5.1: Analytical data of the Id3 polypeptides fragments **V.1-12**.

Product	Sequence	$MW_{calc.}$ (Da)	MW_{found} (Da)	t_r (min.)
V.1	Ac-[K-84]-(48-92)-NH_2	5052.9	5054.6	20.4^a
V.2	Ac-[K(FAM)-84]-(48-92)-NH_2	5410.9	5412.4	23.0^a
V.3	Ac-[K(TAMRA)-84]-(48-92)-NH_2	5465.4	5467.1	23.9^a
V.4	Ac-[K-76]-(48-92)-NH_2	5002.9	5004.5	20.9^a
V.5	Ac-[K(FAM)-76]-(48-92)-NH_2	5360.9	5365.6	25.4^a
V.6	Ac-[K(TAMRA)-76]-(48-92)-NH_2	5415.4	5420.6	25.7^a
V.7	Ac-(48-92)-NH_2	5037.9	5039.6	24.7^a
V.8	Ac-(41-101)-NH_2	6825.9	6824.2	21.3^b
V.9	Ac[K-61]-(31-63)-NH_2	3734.3	3734.1	22.5^c
V.10	Ac-[K(FAM)-61]-(31-63)-NH_2	4092.3	4092.5	24.9^c
V.11	Ac-[K(TAMRA)-61]-(31-63)-NH_2	4146.3	4149.3	24.7^c
V.12	Ac-(31-63)-NH_2	3663.2	3663.9	23.3^c

[a] Gradient: 40% ACN for 5 min., 40-85% ACN over 40 min.

[b] Analytical data previously measured in the Slobodova PhD dissertation [1]; gradient: 10% ACN for 3 min., 10-70% ACN over 40 min.

[c] Gradient: 20% ACN for 5 min., 20-80% ACN over 40 min.

The Id3 peptide analogs were synthesized by solid-phase methodology using the *Fmoc*-chemistry and Rink Amide MBHA resin for peptides **V.1-8** or Rink amide-ChemMatrix resin for peptides **V.9-12**. The introduction of the fluorophore (FAM or TAMRA) at the side chain of lysine was performed on the solid-phase by using HOBt/DIC as coupling reagents. Once the synthesis was completed, cleavage from the solid support and removal of the side chain protecting groups were performed simultaneously by treatment with TFA in the presence of 10% scavengers. Finally, crude products were purified by semi-preparative RP-HPLC and characterized by analytical RP-HPLC and MALDI-TOF mass spectrometry (Table 5.1).

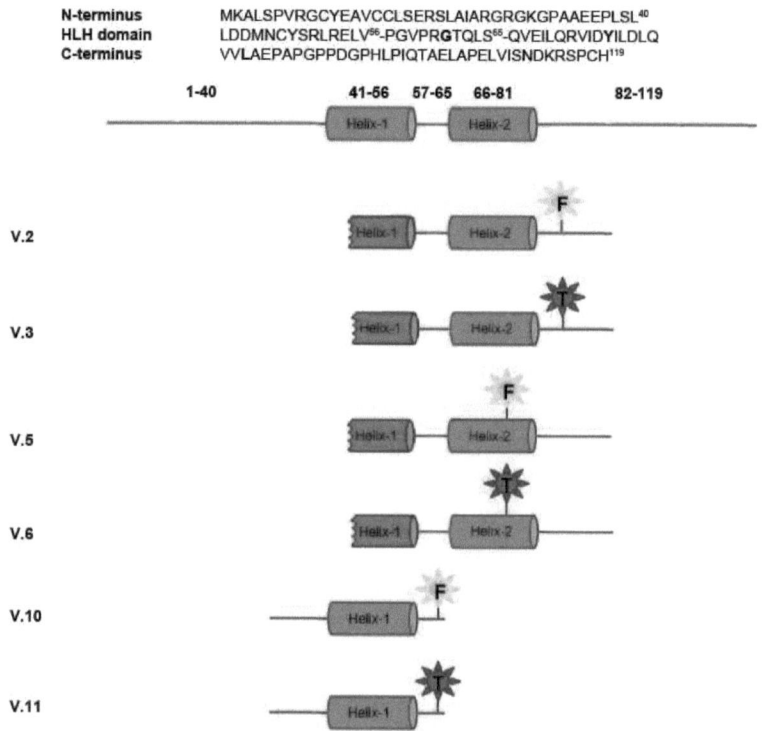

Figure 5.1: Sequence and schematic representation of the Id3 protein domains (N-terminus, HLH motif, C-terminus) and of the fluorescently labeled fragments used in this study (the labeled positions are indicated in bold).

5.3 Results

5.3.1 Conformational studies by CD spectroscopy

The secondary structure of the synthesized Id3 peptides was studied by CD spectroscopy in phosphate buffer (100 mM, pH 7.1). As anticipated, the N-terminally truncated HLH segment 48-92 adopts a less stable helical fold than the complete HLH domain. This is shown by the CD curve of peptide **V.7**, which is still reminiscent of a helix-like motif, but less well-defined, as suggested by the more intense band at 208 nm than that at 222 nm (Figure 5.2A). The CD spectrum of the K/L84 mutant (**V.1**) is characterized by a similar CD shape, but it is more intense than that of peptide **V.7**. This might indicate that the presence of lysine in place of leucine reduces the formation of irregular oligomers, due to higher solubility of the peptide chain. After labeling of K84 with FAM (peptide **V.2**) or TAMRA (peptide **V.3**), the CD spectra were recorded and compared with native segment **V.7**. Interestingly, the CD curve of the FAM-labeled peptide **V.2** was similar to that of the native segment **V.7**, indicating that K(FAM)84 is sufficiently tolerated. However, a blue shift of the band from 208 nm to 205 nm and the loss of positive ellipticity below 200 nm are indicative of partial conformational transition from α-helix to 3_{10}-helix. The TAMRA-labeled peptide **V.3** showed a CD spectrum similar to that of the FAM-labeled one, but with reduced intensity. This is probably due to the more hydrophobic character of the label, which may lead to some peptide aggregation. The K/Y76 mutation (peptide **V.4**) seems to produce a partial α-helix/3_{10}-helix transition, as suggested by the pronounced and slightly blue-shifted band from 208 nm to 205 nm (Figure 5.2B). Such transition becomes more evident after the labeling of K76 with FAM (**V.5**) or TAMRA (**V.6**). This result is not surprising, as K76 is located in the helix-2 fragment, approximately one helix turn far away from the C-end of the helix. Therefore, the mutation of this position is expected to partly destabilize the α-helix, leading to a shorter and/or 3_{10}-helix. Furthermore, it should be noted that FAM and TAMRA have comparable effects on the conformation of the peptide, as indicated by the superposition of the CD curves corresponding to peptides **V.5** and **V.6** (Figure 5.2B).

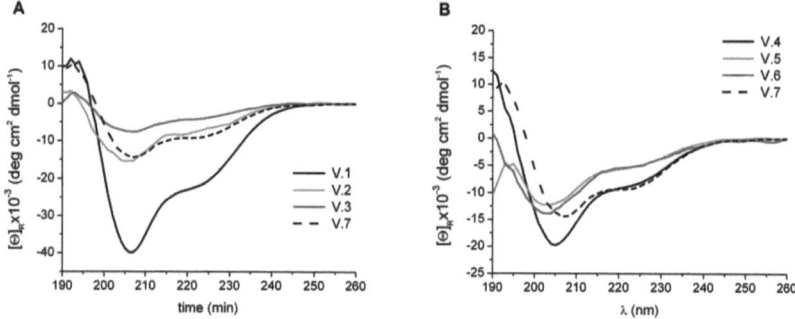

Figure 5.2: CD spectra of the unlabeled, fluorescence-labeled and native Id3 fragments 48-92, modified at position 84 (**A**) or 76 (**B**) in phosphate buffer (100 mM, pH 7.1).

The Id3 segment 31-63 (**V.12**), which contains the helix-1 motif 41-56, mostly adopts a flexible conformation, as indicated by the strong negative CD band at 200 nm and the weak shoulder at 222 nm (Figure 5.3). This is in agreement with the observation of Svobodova [1] that the helix-1 has lower intrinsic helix stability than the helix-2. The K/G61 mutation in the loop (**V.9**) has almost no effect on the CD curve. Also the FAM- or TAMRA-labeled analogs (peptide **V.10** and **V.11**) maintain a flexible conformation. However, especially in the case of the TAMRA-labeled one, a decrease in negative intensity accompanied by a red shift of the band from 200 nm to 203 nm is detected, which may suggest some peptide aggregation, as already observed in the truncated HLH peptide containing TAMRA at position 84.

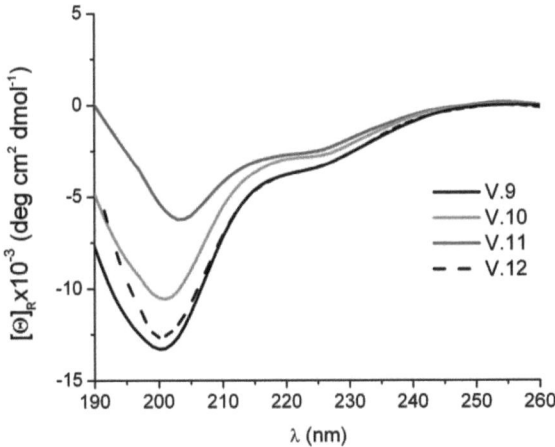

Figure 5.3: CD spectra of the unlabeled, fluorescence-labeled and native Id3 fragments 31-63 modified at position 61 in phosphate buffer (100 mM, pH 7.1).

5.3.2 Fluorescence spectroscopy

The labeled peptides were studied by fluorescence spectroscopy in phosphate buffer (100 mM, pH 7.1) at 5 μM concentration. The emission spectra were recorded upon excitation of FAM at 488 nm at 20°C in the case of the FAM-labeled peptides and in the equimolar mixture of FAM-labeled and TAMRA-labeled peptides. In the case of the TAMRA-labeled peptides alone the emission spectra were recorded upon excitation of TAMRA at 550 nm. The spectra of the FAM-labeled peptides were characterized by a broad band centered at 522 nm and the spectra of the TAMRA-labeled peptides were characterized by a broad band centered at 580 nm (Figure 5.4).

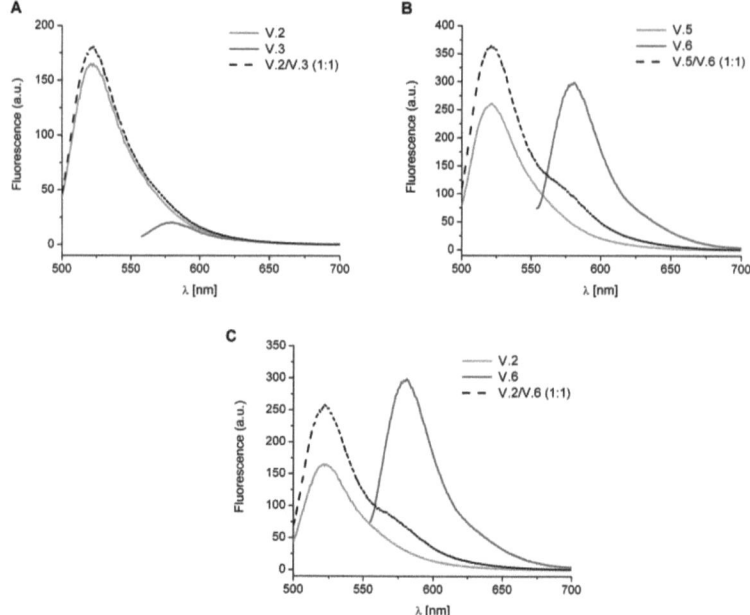

Figure 5.4: Fluorescence spectra of peptides **V.2**, **V.3** (each 5 μM) and equimolar mixture of **V.2** and **V.3** (each 5 μM) (**A**), peptides **V.5**, **V.6** (each 5 μM) and equimolar mixture of **V.5** and **V.6** (each 5 μM) (**B**), peptides **V.2**, **V.6** (each 5 μM) and equimolar mixture of **V.2** and **V.6** (each 5 μM) (**C**).

The TAMRA label located at position 84 of the segment 48-92 (**V.3**) was found to emit weakly compared to when located at position 76 (**V.6**). This reflects the slightly different conformation of the two analogs, as already indicated by the corresponding CD curves. In particular, peptide **V.3** is likely to self-aggregate stronger than peptide **V.6**, which might cause self-quenching of TAMRA. The significant aggregation of **V.3** might also explain, why no detectable FRET was observed for the equimolar mixture **V.2/V.3** (Figure 5.4A). Instead, the equimolar mixture **V.5/ V.6** resulted to be FRET active, as indicated by the appearance of a shoulder around 580 nm, which indicates that there was resonance energy transfer from FAM label of **V.5** to the TAMRA label of **V.6** (Figure 5.4B). FRET activity was also observed for the mixture **V.2/V.6**, which suggests, that position 76 of the one chain and position 84 of the other chain are close enough in the complex to allow resonance energy transfer from FAM at position 84 to TAMRA at position 76 (Figure 5.4C). As already pointed out for the TAMRA emission after direct excitation, also the emission of FAM after excitation differs in intensity, depending on its position in the sequence (76 or 84) as well as on the composition of the complex (**V.5/V.6** or **V.2/V.6**). The FAM emission increases in the following order: **V.2** < **V.5** < **V.2/V.6** < **V.5/V.6**. One of the possible reasons for the increase in the FAM emission, including the FRET-active mixtures, might be a decrease in FAM self-quenching. Another reason might be the increase in folding stability. Indeed, it is known, that polar groups and solvents exert a quenching effect on fluorescence emission. As a result, a more compact fold, in which the fluorescence label is buried in a hydrophobic core, favors strong emissions.

The results of fluorescence studies of peptides **V.10** and **V.11** are shown in Figure 5.5. Again, a weak TAMRA emission after direct excitation was detected for peptide **V.11**, which was attributed to peptide aggregation, as already suggested by the CD curve, and similarly to the behavior of the TAMRA-labeled peptide **V.3**. Moreover, no FRET activity was detected for the mixture **V.10/V.11**, possibly due to the strong self-aggregation of peptide **V.11**.

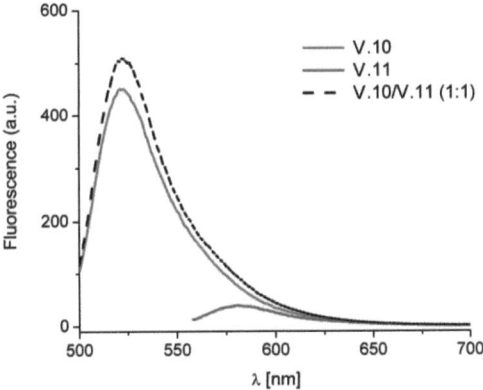

Figure 5.5: Fluorescence spectra of peptides **V.10**, **V.11** (each 5 μM) and equimolar mixture of **V.10** and **V.11** (each 5 μM).

In contrast, the mixtures **V.10/V.3** and **V.10/V.6** were both FRET active (Figure 5.6A, B), which indicates that nascent helix-1 interacts with nascent helix-2. Particular attention should be paid on the FRET-active mixture **V.10/V.3**: indeed, whereas peptide **V.3** was not interacting with peptide **V.2**, it interacts with **V.10**. This suggests, that the interaction of helix-1 with helix-2 is much stronger and more favorable than the helix-2 self-interaction and is thus able to disaggregate peptide **V.3**. Based on this observation, the mixture **V.2/V.11** and **V.5/V.11** were expected to be FRET-active as well. However, a clear shoulder around 580 nm could not be detected (Figure 5.6C, D). Nevertheless, the emission spectrum of the mixtures showed some peculiarities, which support the occurrence of interactions between the two peptides. In particular, the FAM emission of **V.2** and **V.5** in the presence of **V.11** was red-shifted and increased in intensity. Therefore, it might be, that the FAM-labeled peptides are able to weakly interact with the self-aggregates of the TAMRA-labeled peptide **V.11**, without completely disaggregating the latter.

Further, it was observed that the FAM emission at position 61 increased in the following order: **V.10** = **V.10/V.6** < **V.10/V.3**. As already discussed earlier, the reason for that might be the decrease in self-quenching and/or solvent-mediated quenching.

Interestingly, the FAM emission resulted to be higher for all the FRET-active mixtures examined than for the FAM-labeled peptides alone.

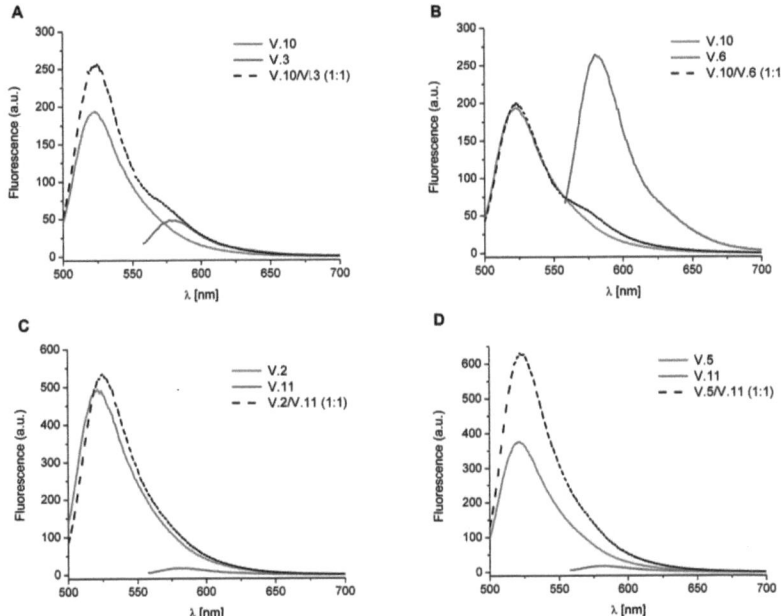

Figure 5.6: Fluorescence spectra of peptides **V.10**, **V.3** (each 5 μM) and equimolar mixture of **V.10** and **V.3** (each 5 μM) (**A**), peptides **V.10**, **V.6** (each 5 μM) and equimolar mixture of **V.10** and **V.6** (each 5 μM) (**B**), peptides **V.2**, **V.11** (each 5 μM) and equimolar mixture of **V.2** and **V.11** (each 5 μM) (**C**), peptides **V.5**, **V.11** (each 5 μM) and equimolar mixture of **V.5** and **V.11** (each 5 μM) (**D**).

5.3.3 Self-recognition between partially and fully folded HLH domains

To see, whether a peptide containing the complete HLH domain and adopting a folded structure (peptide **V.8**, fragment 41-101 of Id3) is able to interact with a peptide reproducing a truncated and partially folded region of the HLH domain (peptide **V.2**), we performed the titration of the fluorescence analog **V.2** with the non fluorescence analog **V.8**. The titration was followed by fluorescence spectroscopy in phosphate buffer (100 mM, pH 7.1) at 10 μM concentration of **V.2**. The emission spectra were recorded upon excitation of FAM at 488 nm at 20°C after each addition of small amounts of analog **V.8**. The spectra of analog **V.2** show a broad band centered at 522 nm, whose fluorescence intensity slightly increases upon addition of analog **V.8** (Figure 5.7). This suggests a change of the environment around FAM and is indicative of an interaction between the two analogs.

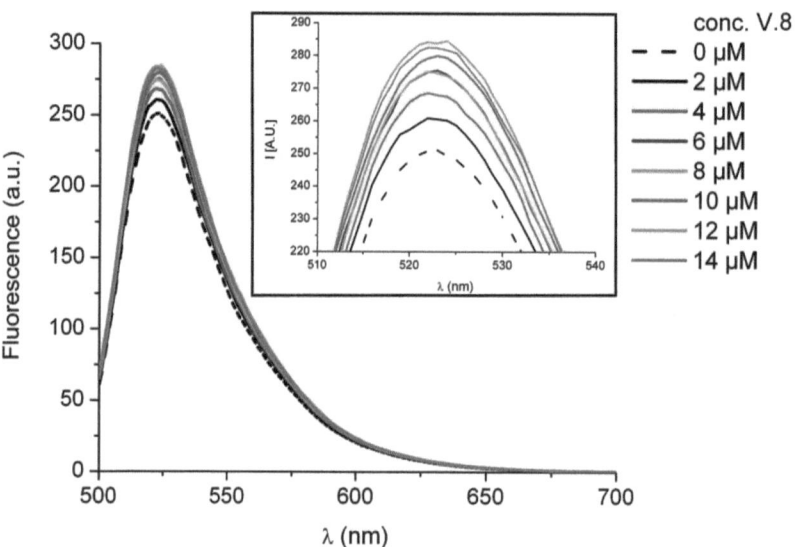

Figure 5.7: Fluorescence titration of peptide **V.2** with **V.8**.

The maximum intensity of emission at 522 nm was plotted against the concentration of analog **V.8** and the result is shown in Figure 5.8. The fitting of the points was made by using the Hill1 equation and the OriginLab software. The obtained sigmoid shows a plateau starting around 10 μM concentration of ligand **V.8**. The extrapolated Hill coefficient n was of 1.8 and the $k(0.5)$ value, which corresponds to the ligand concentration at which half of the binding sites are occupied was of 3.6 μM. The Hill coefficient of about 2 may be indicative of the interaction of **V.2** with the homodimeric form of **V.8**, the latter displaying thus the HLH fold. Moreover, due to the fact that **V.2** contains helix-2 and only a portion of helix-1, its interaction with **V.8** must be mediated by the helix-2 alone.

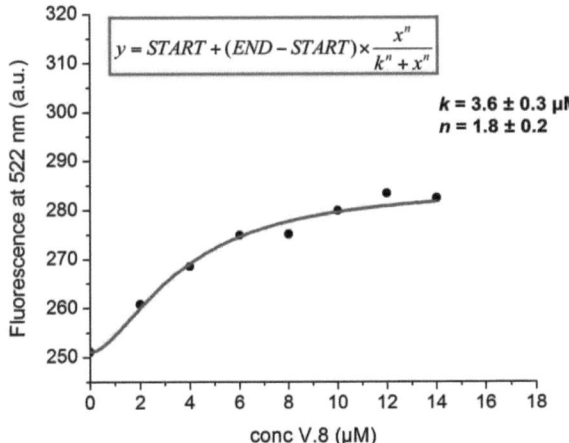

Figure 5.8: Isotherm of the binding of **V.8** to **V.2**, as obtained by fluorescence titration of **V.2** with **V.8**.

5.4 Conclusions

The use of a N-truncated Id3 HLH segment like 48-92 has allowed reproducing a partially unfolded conformer of the HLH domain, in which the helix-2 is assumed to be a locally ordered structure, whereas the preceding and following residues are expected to be flexible. The two corresponding fluorescence labeled analogs were displaying the FAM donor or the TAMRA acceptor either at position 84, beyond the helix-2, or at position 76 within helix-2. These labels did not induce dramatic conformational changes, although in the case of position 84 the TAMRA-labeled triggered peptide aggregation. FRET measurements confirmed the ability of helix-2 to self-associate.

The Id3 fragment 31-61 reproducing the helix-1 and short flanking regions, including the loop, was quite flexible, which confirms the superior helix-2 over the helix-1 stability. The two corresponding FAM or TAMRA labeled analogs at position 61 maintained high flexibility, although, again, the TAMRA acceptor triggered aggregation. FRET measurements between the labeled segments 48-92 and 31-63 confirmed helix-1/helix-2 hetero-association. Moreover, the aggregated TAMRA-labeled helix-2-containing fragments were partially disaggregated by the FAM-labeled helix-1- but not helix-2-containing fragments. This might indicate stronger interaction between helix-1 and helix-2 rather than between two helix-2 fragments.

In conclusion, these results underline the fact that helix-2 is partly folded also in the absence of helix-1. This is not the case of helix-1 that requires the presence of helix-2 to fold. Therefore, helix-2 might represent an early local structure that may trigger protein oligomerization by self-association. This might favor helix-2 contacts with the still poorly defined helix-1, thus leading to the stabilization of the latter and the formation of the four-helix bundle.

References

[1] Svobodova J., PhD Dissertation, University of Regensburg, **2006**.

[2] J. Svobodova, and C. Cabrele, Stepwise Solid-Phase Synthesis and Spontaneous Homodimerization of the Helix-Loop-Helix Protein Id3. *Chem Bio Chem* **2006**, 7, 1164-68.

Chapter 6

Salt and solvent effects on the conformational stability of the Id HLH domain

6.1 Introduction

The most common approach to modulate the conformation and stability of native peptides and proteins is the application of chemical tools, including (a) the replacement of conserved amino-acid residues with other proteinogenic amino acids or non-natural amino acid analogs, and (b) the substitution of peptide-backbone amides with surrogates. Such a chemical approach has been used by the Cabrele group to study and modulate the conformational properties of the HLH domain of the Id proteins [1–8].

All perturbations of the native HLH fold produced by using the chemical approach have been intrinsically induced by modifications of the primary structure. In the present study we have investigated, how extrinsic factors, such as the presence of a cosolvent, cosolute, or temperature, affect the conformational properties of the Id HLH domain. In fact, besides the intrinsic structural propensities dictated by the primary structure, also the environment can easily influence the structure and stability of a protein. Due to the importance of proteins in biology and their utilization in a variety of scenarios going from the formulation of drugs to electrochemical, bio- and nanotechnological applications, it is worth to investigate their behavior upon physical and chemical changes of their environment. The HLH domain is a valuable candidate for a model study, as it possesses the capability of folding into highly helical, non-covalent bundles stabilized by intra- and intermolecular helix-helix interactions, thus containing secondary, tertiary as well as quaternary structure elements. Moreover, the role of the HLH domain in Nature is not limited to the building of tight protein cores, but it extends to the molecular recognition of proteins, which is the mechanism of action of the HLH transcription factors.

Herein, we have evaluated the effect of choline dihydrogenphosphate (chol-dhp), choline chloride (chol-Cl) and 2,2,2-trifluoroethanol (TFE) on the stability of the non-covalent helix bundle of the Id3 and Id4 HLH domains. Chol-dhp, which belongs to the class of salts referred to as ionic liquids and consists of a chaotropic cation and a kosmotropic anion, has been shown to be compatible with the native structure and activity of proteins and, moreover, to allow the protein storage in solution at room temperature for long periods [9]. Further, the presence of chol-dhp has been found to improve the thermal stability of proteins and to prevent the irreversible formation of aggregates. In comparison to chol-dhp, chol-Cl has shown minor stabilizing effect [10].

TFE is known to act as α-helix stabilizing and tertiary structure destabilizing cosolvent [11]. The mechanism of action is still unclear, but it is likely that TFE favors the formation of intramolecular over solute-solvent hydrogen bonds. In addition, due to the presence of the trifluoromethyl group, TFE is able to interact with apolar residues, thus weakening the specific hydrophobic interactions that are determinant for the protein to be folded [12].

6.2 Results

6.2.1 Chol-dhp negatively affects the helical content of the Id3 HLH domain

The CD spectrum of the Id3 HLH domain in phosphate buffer (100 mM, pH 7) is characteristic of an α-helix-rich structure, as indicated by the strong positive band at 194 nm and the two minima at 220 nm and 208 nm (Figure 6.1A). The ellipticity ratio R between the two minima is 0.914. Deconvolution of the CD spectrum by using the Contin algorithm gave the following secondary structure fractions: 46% helix, 12% strand, 17% turns, and 25% disordered. The shape of the curve changes significantly by increasing the concentration of chol-dhp up to 4 M, and a measurable decrement of the intensity of the three bands could be detected, which indicates a destabilization of the helix fold (Figure 6.1A). The change of the R value at different salt concentrations was sigmoidal, with an inflection point around 2 M chol-dhp, which suggests a cooperative process with a two-state transition supported by the presence of the isodichroic point at 199 nm. No precipitation was detected by addition of chol-dhp and this suggests that the decrement of intensity of the three bands could be related to the formation of soluble aggregates. Indeed, the results from the deconvolution of the curves (Table 6.1) shows a transition from α-helix structure to β-strand, as indicated by the strand fraction increasing from 12% to 23% at the expense of the helix fraction that decreased from 46% to 31%. We believe that chol-dhp could induce partial peptide aggregation and these aggregates are supposed to adopt ill-defined structures, including β-sheet-like elements.

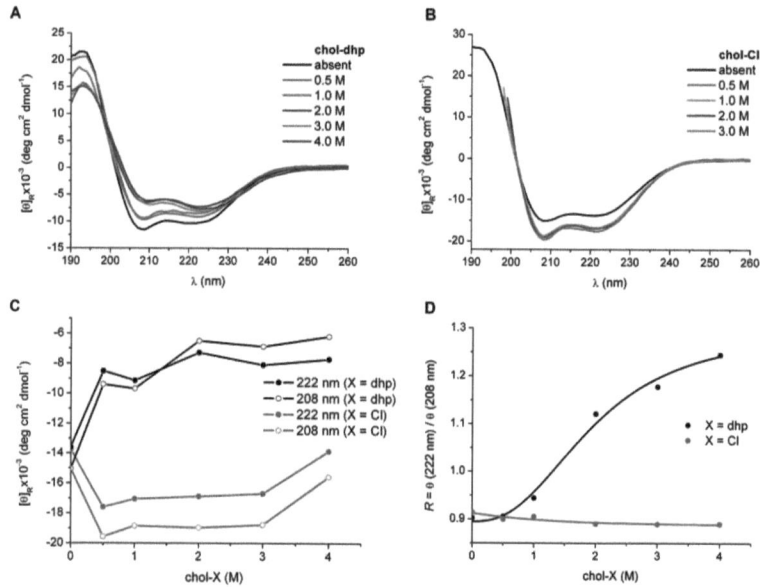

Figure 6.1: CD spectra of the Id3 HLH domain (40 μM in 100 mM phosphate buffer, pH 7) at different concentrations of chol-dhp (**A**) and chol-Cl (**B**). Dependence of the minima at 222 nm and 208 nm (**C**), and of their ratio R (**D**) on chol-dhp/Cl concentration.

Table 6.1: Secondary structure elements composition of the Id3 HLH domain (the CD spectra deconvolution was performed by using the algorithm Contin [13] from the online server Dichroweb [14]).

[chol-dhp] (M)	T (°C)	Helix (%)	Strand (%)	Turn (%)	Disordered (%)
-	20	46	12	17	25
0.5	20	33	21	21	27
1	20	35	21	17	27
2	20	31	22	21	26
3	20	31	23	19	26
4	20	31	23	19	26

6.2.2 Chol-Cl positively affects the helix bundles of the Id3 HLH domain

The intensity of the CD spectrum of the Id3 HLH domain, unlike the Id4 HLH, in the presence of chol-Cl increased, which indicates a stabilization of the helix fold (Figure 6.1B). However, the shape of the CD curve did not change significantly, as indicated by the fact that the R value remained almost constant (Figure 6.1D). Unfortunately, the CD spectra could not be deconvoluted, due to the missing experimental data below 200 nm. Indeed, the high concentration of Cl^- prevented the recording of the spectra until 190 nm. Despite the lack of deconvultion, it is assumed, that the increased CD intensity reflects an increase in helix content probably at the expense of higher-order oligomers containing β-sheets-like structures.

6.2.3 Chol-dhp favors tight and highly structured helix bundles of the Id4 HLH domain

The CD spectrum of the Id4 HLH domain in phosphate buffer (100 mM, pH 7) is characteristic of an α-helix-rich structure, as indicated by the strong positive band at 194 nm and the two minima at 220 nm and 208 nm (Figure 6.2A). The ellipticity ratio R between the two minima is 0.915, practically the same value of the Id3 HLH domain (R = 0.914). Deconvolution of the CD spectrum by using the Contin algorithm gave the following secondary structure fractions: 53% helix, 4% strand, 18% turns, and 25% disordered (Table 6.2). Upon increasing the concentration of chol-dhp up to 4 M, the shape of the curve did not change significantly, although a measurable increment of the intensity of the three bands could be detected. The increase in ellipticity is more pronounced at 220 nm rather than at 208 nm (Figure 6.2C), with a consequent linear increase of the R value from 0.915 up to 0.973 (Figure 6.2D). We have attributed these CD changes to a positive effect of chol-dhp on the helix packing, which consequently enhanced the helix content: indeed, the helix fraction increases from 53% to 60% at the expense of the disordered fraction that decreased from 25% to 19%, whereas the strand and turns fractions remained unchanged (Table 6.2). Such two-state transition is supported by the presence of the isodichroic point at 206 nm (Figure 6.2A).

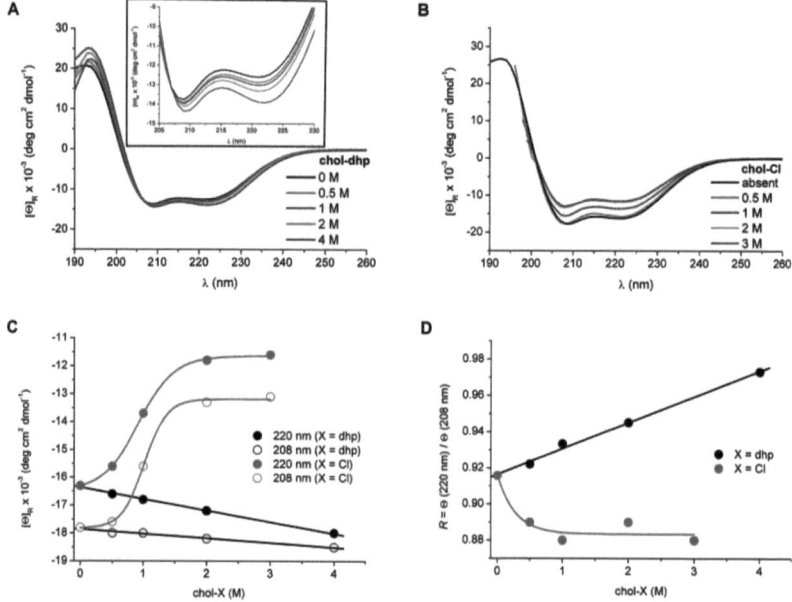

Figure 6.2: CD spectra of the Id4 HLH domain (90 µM in 100 mM phosphate buffer, pH 7) at different concentrations of chol-dhp (**A**) and chol-Cl (**B**). Dependence of the minima at 220 nm and 208 nm (**C**), and of their ratio R (**D**) on chol-dhp/Cl concentration.

Table 6.2: Secondary structure elements composition of the Id4 HLH domain (the CD spectra deconvolution was performed by using the algorithm Contin [13] from the online server Dichroweb [14]).

[chol-dhp] (M)	TFE (%)	T (°C)	Helix (%)	Strand (%)	Turn (%)	Disordered (%)
-	-	20	53	4	18	25
0.5	-	20	56	4	18	22
1	-	20	58	4	20	18
2	-	20	59	4	18	19
4	-	20	60	4	17	19
-	-	85	26	18	25	31
0.6	-	85	58	4	20	18
1.3	-	85	58	4	20	18
-	-	20	53	4	18	25
-	10	20	50	8	17	25
-	20	20	50	8	17	25
-	40	20	55	4	17	24
-	50	20	55	3	18	24
-	80	20	61	3	16	20
1.3	-	20	58	4	20	18
1.3	-	20	50	8	16	26
1.3	-	20	53	6	17	24
1.3	-	20	57	3	17	23
1.3	-	20	59	3	16	22

6.2.4 Chol-Cl negatively affects the helical content of the Id4 HLH domain

The intensity of the CD spectrum of the Id4 HLH domain in the presence of chol-Cl decreased, which indicates a destabilization of the helix fold (Figure 6.2B). The change of the intensity of the two negative bands at 208 nm and 220 nm at different salt concentrations was sigmoidal, with an inflection point at 1 M chol-Cl, which suggests a specific effect of chol-Cl on the conformation (Figure 6.2C). Deconvolution of the CD spectra could not be performed, as the presence of high concentrations of Cl^- prevented CD measurements below 197 nm. However, based on the fact that the CD curves differ in intensity but not significantly in shape, we believe that chol-Cl induced partial peptide aggregation, probably by weakening electrostatic repulsions through the counterion action of Cl^- on the basic residues. These aggregates are supposed to adopt ill-defined structures, including β-sheets-like elements.

6.2.5 Effect of chol-dhp and chol-Cl on the thermal dissociation of the helix bundles

Id3 HLH thermal dissociation. The thermal stability of the Id3 HLH domain in phosphate buffer was investigated over the temperature range from 20 °C to 90 °C. The curve obtained by reporting the ellipticity values at 222 nm versus the temperature is described by an asymmetrical sigmoid with an inflection point at 66 °C (Figure 6.3). In the presence of 1 M chol-Cl the inflection point was shifted to 60 °C. This moderate decrease in thermal stability in the presence of the salt may reflect the absence of oligomers containing β-sheet-like structures, which usually dissociate at higher temperatures than α-helical folds. This is supported by the observation that the T_m value of the peptide in the presence of 1 M chol-dhp, which was shown by CD spectroscopy to contain aggregates based on β-sheets, was around 75 °C. After heating, the peptide solutions were cooled down, and the native fold was fully recovered only in absence of salt and in the presence of 1 M chol-Cl, indicating that the partial thermal denaturation was reversible (data not shown). Interestingly, in the presence of 1 M chol-dhp the thermal denaturation was not reversible.

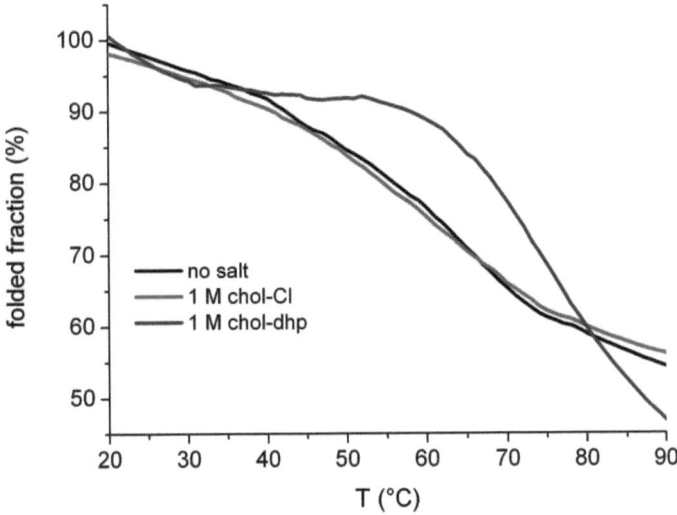

Figure 6.3: Thermal denaturation of the Id3 HLH domain (40 μM in 100 mM phosphate buffer, pH 7) in the presence of chol-dhp/Cl.

Id4 HLH thermal dissociation. The thermal stability of the Id4 HLH domain in phosphate buffer was investigated over the temperature range from 20 °C to 85 °C. The curve obtained by reporting the ellipticity values at 220 nm versus the temperature is described by an asymmetrical sigmoid with an inflection point at 72 °C (Figure 6.4). This value is higher than the value of Id3 HLH, which indicates that the helix bundle of Id4 is thermodynamically more stable. In contrast, no inflection point was detected for the heat curves obtained in the presence of 0.6 M and 1.3 M chol-dhp, which indicates a shift of the T_m value above 85 °C. Therefore, chol-dhp confers remarkable thermal stability to the helix bundles. As a matter of fact, deconvolution of the CD spectra recorded at 85 °C in the absence and in the presence of 0.6 M or 1.3 M chol-dhp showed that 26%, 31% and 39% helix fraction was still present, respectively. After heating, the peptide solutions were cooled down, and the native fold was fully recovered, indicating that the thermal denaturation was reversible (data not shown).

Interestingly, also the presence of chol-Cl increased the T_m value of the helix bundles, however, at a remarkably minor extent in comparison to chol-dhp: indeed, the T_m values corresponding to 0.5 M and 1 M chol-Cl are 75 and 76 °C, respectively (Figure 6.4). This moderate shift of T_m towards higher values probably reflects the presence of aggregates with ill-defined structures, including β-sheets-like elements, which dissociate at higher temperatures. Indeed, β-sheets are known to unfold at higher temperatures than α-helices [15].

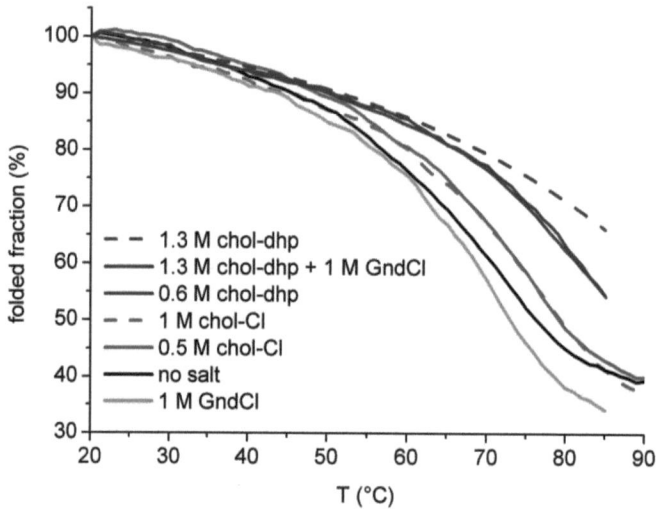

Figure 6.4: Thermal denaturation of the Id4 HLH domain (60 μM in 100 mM phosphate buffer, pH 7) in the presence of chol-dhp/Cl and GndCl.

6.2.6 Chol-dhp-stabilized helix bundles are resistant to the denaturing effect of GndCl

The behavior of the Id4 HLH domain upon denaturing conditions was investigated in phosphate buffer containing GndCl at different concentrations. As shown in Figure 6.5A and D, no loss of helical conformation was detected below 5 M GndCl. Only at higher concentrations of the denaturant there was a partial destabilization of the structure. However, the loss of the helix fraction at the maximal GndCl concentration used (7.2 M) was limited to 34%. Instead, the Id3 HLH domain shows loss of helical conformation at concentrations of the denaturant GndCl higher than 2 M, whith complete unfolding at 6 M GndCl, as presented in the dissertation work of Svobodova [8]. Therefore, the Id3 HLH domain results to be less resistant to the denaturing effect of GndCl than the Id4 HLH domain. This is also in agreement with the lower T_m value of the Id3 than that of the Id4 HLH domain.

In the presence of chol-dhp the Id4 HLH domain was almost completely unaffected by increasing amounts of GndCl (Figure 6.5B and D): indeed, only 15% loss of helix fraction was observed at 7.2 M GndCl.

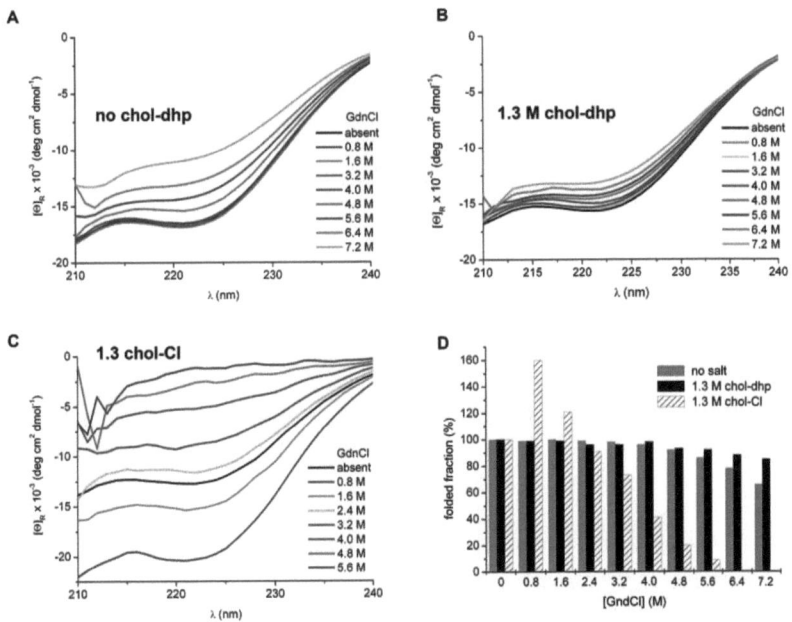

Figure 6.5: Titration of the Id4 HLH domain (50 μM in 100 mM phosphate buffer, pH 7) with GndCl in the absence (**A**) and in the presence of 1.3 M chol-dhp(**B**) or chol-Cl (**C**). Variation of the folded fraction (based on the ellipticity value at 220 nm) at increasing GndCl concentrations is reported in panel **D**.

Since GndCl does not apparently affect the overall conformation of the Id4 HLH bundle below 5 M concentration, we also investigated the effect of GndCl on the thermal stability. In the presence of 1 M GndCl, the thermal denaturation process was more cooperative, with only a slight destabilization of the fold (71 °C versus 72 °C). A similar behavior was observed for the chol-dhp-stabilized Id4 HLH domain in the presence of 1 M GndCl (Figure 6.4).

6.2.7 Chol-Cl strengthens the denaturing effect of GndCl on the helix bundles

The GndCl titration of the Id4 HLH domain in the presence of 1.3 M chol-Cl was completely different from that in the presence of 1.3 M chol-dhp (Figure 6.5C, D). Indeed, upon addition of 0.8 M GndCl the intensity of the CD band at 220 nm increased of about 60%, resembling the highest one shown in the CD spectrum of the Id4 HLH domain in 4 M chol-dhp. However, at higher concentrations of the denaturant the folded fraction decreased quickly, and the HLH domain was completely unfolded at 5.6 M GndCl. The stabilization of the helical fold at low denaturant concentration probably reflects the disrupting effect of GndCl on the chol-Cl-induced aggregates, thus promoting their refolding into a helix-rich conformation. However, this conformation seems to be much more sensitive to the denaturing effect of GndCl than the chol-dhp-stabilized and even salt-free ones. This indicates that GndCl-induced unfolding of the helix bundle can be triggered by the presence of choline, but counteracted by the presence of the dhp anion.

6.2.8 Chol-dhp supports the TFE-induced unbundling of the helix bundles

TFE is known to stabilize helical structures of peptides and polypeptides by "dehydrating" the backbone, thus promoting the formation of intramolecular hydrogen bonds. However, the fluorinated alcohol is also known to destabilize tertiary and quaternary structures by competing with the apolar amino-acid side chains involved in hydrophobic interactions. This implies that TFE-promoted helices are monomeric. Therefore, TFE titration of the Id4 HLH domain was expected to induce a transition from packed to single helices. In fact, increase in the TFE/buffer ratio up to 1:1 was accompanied by minor changes of the minimum at 220 nm, but by a significant increase in the intensity of the minimum at 208 nm (Figure 6.6A, C). As a result, the R value showed a sigmoidal decrease with an inflection point close to 15% TFE, indicating the dissociation of the helix bundle (Figure 6.6D). Deconvolution of the CD spectra of the TFE-containing peptide samples revealed a small increase in the β-sheet fraction at the expense of the helix fraction with 10-20% TFE compared to the TFE-free sample (α/β = 50:8 versus 53:4). However, the helix content was recovered at higher TFE percentages (α/β = 55:3 by 50% TFE).

A similar behavior was observed upon TFE titration in the presence of 1.3 M chol-dhp (Figure 6.6B, C); however, the dissociation of the helix bundle appeared to occur already below 15% TFE, as suggested by the shape of the curve of the R values (Figure 6.6D). At 15% TFE the secondary structure elements composition matches that found for the salt-free samples containing 10-20% TFE (α/β/turns/disordered = 50:8:16:26). This indicates that the salt has no effect on the conformation. The unbundling appears to be more cooperative in the presence rather than in the absence of chol-dhp (Figure 6.6D). This, together with the early dissociation of the helix bundle, may implicate a role of the salt, in particular of choline, whose hydrophobic character can additionally favor the appearance of single helices.

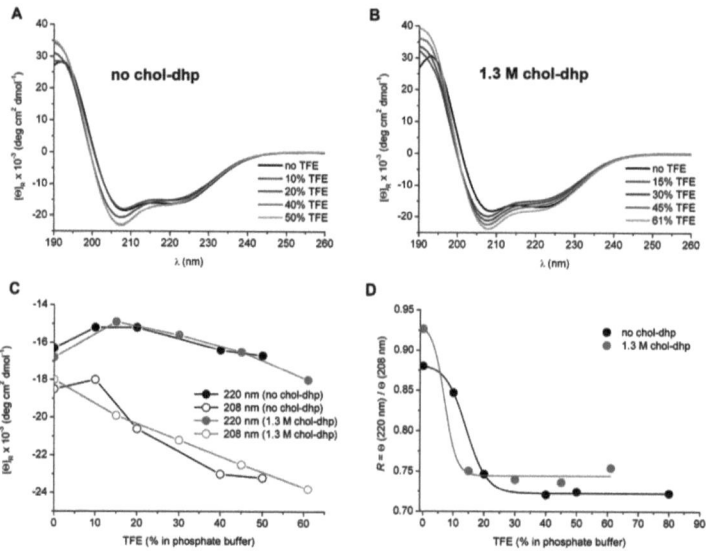

Figure 6.6: CD spectra of the Id4 HLH domain (61 μM in 100 mM phosphate buffer, pH 7) upon TFE titration in the absence (**A**) and in the presence (**B**) of chol-dhp. Dependence of the minima at 220 nm and 208 nm (**C**), and of their ratio R (**D**) on chol-dhp concentration.

6.3 Discussion

Proteins containing the HLH domain often are homo- and/or heterodimerized. For example, the bHLH transcription factors are able to bind to the DNA only as dimers. Although their biologically relevant state is the dimeric one, nevertheless the bHLH domains of some proteins, including MyoD and myogenin, are also found as tetramers. The Id proteins act as inhibitors of DNA binding by interacting with the bHLH factors and forming heterodimers which are devoid of the ability to complex the DNA. Like the bHLH MyoD domain, also the Id HLH domain was found to prefer the tetrameric over the dimeric state [16]. By analyzing the amphipathicity of the two helices of the Id3 and Id4 HLH domains, it is evident that the hydrophobic face of helix-1 is less extended than the hydrophilic one. The opposite holds for helix-2, whose hydrophilic face is strongly negatively charged because of the presence of one glutamate and two aspartates (Figure 6.7). Due to its broad hydrophobic surface, helix-2 is believed to play a key role in the formation and stabilization of the dimers, and especially of the tetramers, as a major number of interhelical contacts is allowed and even necessary to minimize the solvent accessibility of the apolar side chains.

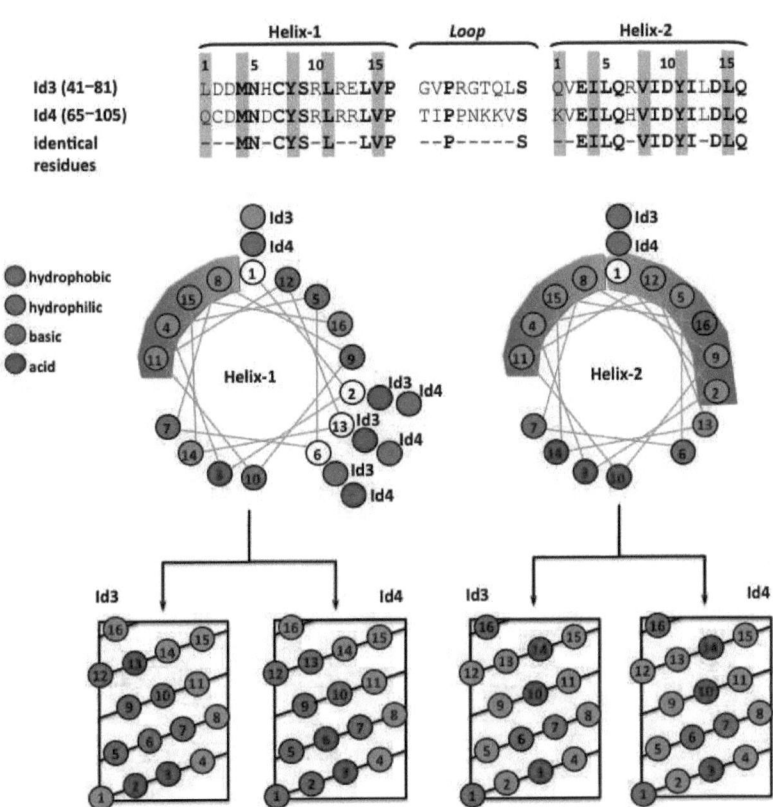

Figure 6.7: Amino-acid sequence of the HLH domain of Id3 and Id4 and helical wheels of helix-1 and helix-2 (the residues in bold represent the conserved residues within the Id protein family).

Since the self-association of the Id HLH domain is driven by hydrophobic interactions that lead to the formation of an amphipatic helix bundle [5], such a process will be sensitive to changes of the apolar character of the environment. For example, increasing the hydrophobicity of the environment will favor the presence of solvated single solute molecules over amphipathic helix bundles. In fact, this is what we observed in the presence of TFE that dramatically weakened the interhelical hydrophobic contacts.

In contrast, the salt GndCl exerted no or rather a positive effect on the self-associated state of the Id4 HLH peptide. GndCl is a well-known denaturant of proteins at high concentrations but shows stabilizing effects at submolar concentrations [17], nevertheless there were no detectable changes of the overall conformation at denaturant concentrations up to 5 M. The stabilizing effect of GndCl at submolar concentrations has been attributed to the guanidinium cation and its ability to bind cation-binding sites present in the folded protein [18]. The remarkable stability of the Id4 HLH bundles against GndCl might also reflect the presence of favorable interactions between the guanidinium ion and the helix bundles. The main contribution of the cation binding is likely to come from the acidic residues on the hydrophilic surfaces of helix-1 and helix-2. In particular, we observed that, within the four Id HLH domains, the Id2 and Id4 HLH domains were significantly more stable against GndCl denaturation than the Id1 and Id3 HLH domains: indeed, the latter were 50% unfolded at 4 M GndCl [8], while the Id4 HLH domain was still fully folded and the Id2 HLH domain displayed an even enhanced helix content. The latter two domains share the presence of an aspartate at position 6 of helix-1, whereas the Id1 and Id3 HLH domains contain a glycine and histidine residue, respectively. Therefore, the negatively charged surface is expected to be larger for the Id2 and Id4 than for the Id1 and Id3 HLH domains.

The stabilizing effect of chol-dhp on the helix bundle was evident in the case of the Id4 HLH domain. Indeed, improved helix packing combined with enhanced helix content at the expense of disordered structure was detected in the presence of increasing chol-dhp concentrations. Moreover, the chol-dhp-stabilized fold displayed superior thermal stability as well as resistance against chemical denaturation. Instead, in the case of the Id3 HLH domain chol-dhp induced a transition from α-helix to β-strand structure.

Although positive effects of hydrated chol-dhp on the thermal stability of folded proteins have been reported [10, 19], the reason for that is not fully understood yet. One hypothesis based on DSC measurements is that chol-dhp should destabilize the partially unfolded intermediates and/or prevent their irreversible transition into poorly soluble aggregates [10]. However, other questions are still open, especially concerning the role of the cation and anion and their mode of action: for example, it has been recently shown that chol-dhp, but not chol-Cl, increased the thermal stability of RNase [10], thus underlining the superior role of the anion over the cation. However, the stabilization of RNase T1 at submolar GndCl concentrations has been attributed to the cation [17]. Moreover, it has been reported that cation-specific effects mediate the salt-induced stabilization of apoflavodoxin [20]. The stabilizing or destabilizing role of the cation or anion may result from direct interactions with potential cation/anion-binding sites displayed by the native structure. For example, molecular dynamics studies carried out by Vrbka et al. on BPTI and HRP have shown that choline can bind to a large variety of side chains (i.e., Asp, Glu, Gln, Arg, Ser, Tyr, Met, Pro, Leu) and is characterized by larger contact times than the sodium cation, whereas Cl^- displays no preference for the protein surface [21]. Also in the case of the thermal stabilization induced by submolar concentrations of GndCl (0-0.3 M) to RNase T1, Mayr and Schmid postulated favorable contacts of the guanidinium ion with the native protein [22].

In this chapter we showed that hydrated chol-dhp and chol-Cl affected the Id HLH fold in a different manner. However, the changes were related to the target peptide too. Indeed, chol-dhp stabilized the helix-rich fold of the Id4 HLH domain, whereas it destabilized the fold of the Id3 HLH domain. In contrast, chol-Cl destabilized the fold of the Id4 HLH domain, whereas it stabilized the Id3 HLH domain. The different behavior between the two HLH domains in the presence of the two salts (chol-dhp and chol-Cl) could be related to the different amino-acid pattern at the hydrophilic face of helix-1: indeed, this displays some differences in Id3 and Id4, such as D2/C2, H6/D6 and E13/R13, whereas the hydrophilic face of helix-2 is highly conserved in both proteins. This different amino-acid pattern might lead to different interactions whit the anion of the salt (dhp^- or Cl^-) and, as a result, to the stabilization of α-helices or β-sheets.

Probably, Cl⁻ acts as counterion of basic residues of the Id4 HLH domain, which might lead to the loss of helix-stabilizing intramolecular salt bridges (i.e., D6-R10). Instead, the positive effect of Cl⁻ on the Id3 HLH domain might be due to the fact that its role as counterion reduces electrostatic repulsions between positive side chains (i.e. H6-R10), which would lead to a stabilization of helix-1.

The Id4 HLH fold has been shown to be highly resistant against GndCl-mediated denaturation. However, the addition of chol-Cl, but not of chol-dhp, has dramatically favored the unfolding process. This is likely to reflect the fact that chol-Cl had shown a destabilizing effect on the Id4 HLH fold, whereas chol-dhp had stabilized it. Therefore, it appears that GndCl and chol-Cl work together to induce the denaturated state. In contrast, the stabilizing effect of chol-dhp appears to fully overcome the denaturing effect of GndCl.

The TFE-induced unbundling of the Id4 HLH fold appears to be supported by the presence of chol-dhp. This is likely to be due to the fact that choline can interact with hydrophobic residues, thus stabilizing the solvent-exposed hydrophobic face of the single helices.

6.4 Conclusions

In this chapter we have investigated the effect of chol-Cl and chol-dhp on the HLH fold of Id3 and Id4. We have found that chol-dhp favors a disordered-to-helix transition of the Id4 HLH domain, but a helix-to-β-sheets transition of the Id3 HLH domain. Instead, chol-Cl favors helix-to-β-sheets transition of the Id4 HLH domain, but the opposite transition in the case of Id3 HLH domain. This suggests that the effect of these two salts on the conformation of a protein depends on the protein itself. With regard to the thermal denaturation, we observed the the T_m increased when the salt favored the β-sheets fraction. This is in agreement with the fact that β-sheets are thermally more stable than helices. Interestingly, the T_m increased after salt-induced helix-content increase at the expense of the random coil fraction, but it moderately decreased when the helix stabilization was at the expense of the β-sheets fraction. Moreover, chol-dhp suppresses the GndCl-induced unfolding of the Id4 helix bundle, but it favors the

TFE-induced unbundling of the helices. Interestingly, the presence of chol-Cl may accelerate the GndCl-mediated denaturation of the Id4 HLH fold. Altogether, these results on the effect of cosolutes on the HLH fold of Id3 and Id4 raise the question of the role of single amino acids or pattern of amino acids in the interplay with the cosolutes. It would be very important to find general rules for the use of salts and solvents to stabilize proteins. However, this might require first deep understanding of the direct and indirect function of salts and solvent molecules on the folding and dynamics of proteins.

References

[1] Beisswenger, M.; Yoshiya, T.; Kiso, Y.; Cabrele, C., Synthesis and conformation of an analog of the helix-loop-helix domain of the Id1 protein containing the O-acyl iso-prolyl-seryl switch motif. *J Pept Sci* **2010**, 16 (6), 303-8.

[2] Kiewitz, S. D.; Kakizawa, T.; Kiso, Y.; Cabrele, C., Switching from the unfolded to the folded state of the helix-loop-helix domain of the Id proteins based on the O-acyl isopeptide method. *J Pept Sci* **2008**, 14 (11), 1209-15.

[3] Svobodova, J.; Cabrele, C., Stepwise solid-phase synthesis and spontaneous homodimerization of the helix-loop-helix protein Id3. *Chembiochem* **2006**, 7 (8), 1164-8.

[4] Colombo, N.; Cabrele, C., Synthesis and conformational analysis of Id2 protein fragments: impact of chain length and point mutations on the structural HLH motif. *J Pept Sci* **2006**, 12 (8), 550-8.

[5] Kiewitz, S. D.; Cabrele, C., Synthesis and conformational properties of protein fragments based on the Id family of DNA-binding and cell-differentiation inhibitors. *Biopolymers* **2005**, 80 (6), 762-74.

[6] Colombo, N. Synthesis and conformational analysis of polypeptides related to the inhibitor of the DNA binding and cell differentiation Id2. University of Regensburg, **2006**.

[7] Kiewitz, S. D. Structural investigations of the Id helix-loop-helix dimerization domain. University of Regensburg, Regensburg, **2007**.

[8] Svobodova, J., PhD Dissertation: Id3, Inhibitor of DNA Binding and Cell Differentiation: Synthesis and Conformational Analysis of the Full-Length Protein and its Truncated Analogues. University of Regensburg, **2007**.

[9] Fujita, K.; Forsyth, M.; MacFarlane, D. R.; Reid, R. W.; Elliott, G. D., Unexpected improvement in stability and utility of cytochrome c by solution in biocompatible ionic liquids. *Biotechnol Bioeng* **2006**, 94 (6), 1209-13.

[10] Constatinescu, D.; Herrmann, C.; Weingartner, H., Patterns of protein unfolding and protein aggregation in ionic liquids. *Phys Chem Chem Phys* **2010**, 12 (8), 1756-63.

[11] Sivaraman, T.; Kumar, T. K.; Yu, C., Destabilization of native tertiary structural interactions is linked to helix-induction by 2,2,2-trifluoroethanol in proteins. *Int J Biol Macromol* **1996**, 19 (4), 235-9.

[12] Luidens, M. K.; Figge, J.; Breese, K.; Vajda, S., Predicted and trifluoroethanol-induced alpha-helicity of polypeptides. *Biopolymers* **1996**, 39 (3), 367-76.

[13] Provencher, S. W.; Glockner, J., Estimation of globular protein secondary structure from circular dichroism. *Biochemistry* **1981**, 20 (1), 33-7.

[14] Whitmore, L.; Wallace, B. A., DICHROWEB, an online server for protein secondary structure analyses from circular dichroism spectroscopic data. *Nucleic Acids Res* **2004**, 32 (Web Server issue), W668-73.

[15] Iloro, I.; Chehin, R.; Goni, F. M.; Pajares, M. A.; Arrondo, J. L., Methionine adenosyltransferase alpha-helix structure unfolds at lower temperatures than beta-sheet: a 2D-IR study. *Biophys J* **2004**, 86 (6), 3951-8.

[16] Fairman, R.; Beran-Steed, R. K.; Anthony-Cahill, S. J.; Lear, J. D.; Stafford, W. F., 3rd; DeGrado, W. F.; Benfield, P. A.; Brenner, S. L., Multiple oligomeric states regulate the DNA binding of helix-loop-helix peptides. *tProc Natl Acad Sci U S A* **1993**, 90 (22), 10429-33.

[17] Mayr, L. M.; Schmid, F. X., Stabilization of a protein by guanidinium chloride. *Biochemistry* **1993**, 32 (31), 7994-8.

[18] Zarrine-Afsar, A.; Mittermaier, A.; Kay, L. E.; Davidson, A. R., Protein stabilization by specific binding of guanidinium to a functional arginine-binding surface on an SH3 domain. *Protein Sci* **2006**, 15 (1), 162-70.

[19] Constantinescu, D.; Weingartner, H.; Herrmann, C., Protein denaturation by ionic liquids and the Hofmeister series: a case study of aqueous solutions of ribonuclease A. *Angew Chem Int Ed Engl* **2007**, 46 (46), 8887-9.

[20] Maldonado, S.; Irun, M. P.; Campos, L. A.; Rubio, J. A.; Luquita, A.; Lostao, A.; Wang, R.; Garcia-Moreno, E. B.; Sancho, J., Salt-induced stabilization of apoflavodoxin at neutral pH is mediated through cation-specific effects. *Protein Sci* **2002**, 11 (5), 1260-73.

[21] Vrbka, L.; Jungwirth, P.; Bauduin, P.; Touraud, D.; Kunz, W., Specific ion effects at protein surfaces: a molecular dynamics study of bovine pancreatic trypsin inhibitor and horseradish peroxidase in selected salt solutions. *J Phys Chem B* **2006**, 110 (13), 7036-43.

[22] O'Toole, P. J.; Inoue, T.; Emerson, L.; Morrison, I. E.; Mackie, A. R.; Cherry, R. J.; Norton, J. D., Id proteins negatively regulate basic helix-loop-helix transcription factor function by disrupting subnuclear compartmentalization. *J Biol Chem* **2003**, 278 (46), 45770-6.

Chapter 7

Experimental part

7.1 Materials

The N^α-$Fmoc$ protected amino acids were purchased from Biosolve (Valkenswaard, the Netherlands), Novabiochem (Merk Biosciences GmbH, Schwalbach/Ts., Germany) and Iris Biotech GmbH (Marktrewitz, Germany):

- $Fmoc$-Ala-OH (MW: 311.3 g/mol)

- $Fmoc$-Cys(Trt)-OH (MW: 585.7 g/mol)

- $Fmoc$-Asp(OtBu)-OH (MW: 411.5 g/mol)

- $Fmoc$-Glu(OtBu)-OH (MW: 425.5 g/mol)

- $Fmoc$-Gly-OH (MW: 297.3 g/mol)

- $Fmoc$-His(Trt)-OH (MW: 619.7 g/mol)

- $Fmoc$-Ile-OH (MW: 353.4 g/mol)

- $Fmoc$-Leu-OH (MW: 353.4 g/mol)

- $Fmoc$-Lys(Boc)-OH (MW: 468.5 g/mol)

- $Fmoc$-Lys(Mtt)-OH (MW: 624.8 g/mol)

- $Fmoc$-Met-OH (MW: 371.5 g/mol)

- *Fmoc*-Asn(Trt)-OH (MW: 596.7 g/mol)
- *Fmoc*-Pro-OH (MW: 337.4 g/mol)
- *Fmoc*-Gln(Trt)-OH (MW: 610.7 g/mol)
- *Fmoc*-Arg(Pbf)-OH (MW: 648.8 g/mol)
- *Fmoc*-Ser(tBu)-OH (MW: 383.4 g/mol)
- *Fmoc*-Thr(tBu)-OH (MW: 397.5 g/mol)
- *Fmoc*-Val-OH (MW: 339.4 g/mol)
- *Fmoc*-Tyr(tBu)-OH (MW: 459.5 g/mol)

Rink 4-methylbenzhydrylamine (MBHA) resin was from Iris Biotech and Rink Amide-ChemMatrix resin from Biotage.

HBTU, HOBt, DIPEA and TFA were from Biosolve (Valkenswaard, the Netherlands); triisopropylsilane (TIS), trimethylsilylbromide (TMSBr), ethanedithiol (EDT) and α-cyano-4-hydroxycinnamic acid were from Fluka (Taufkirchen, Germany).

The following peptide-synthesis-grade reagents were purchased from Biosolve (Valkenswaard, the Netherlands): piperidine, 1-methyl-2-pyrrolidinone (NMP), DMF and diethylether. HPLC-grade acetonitrile (ACN) and TFA for UV spectroscopy were from Biosolve. Sodium hydrogenphosphate and dihydrogenphosphate were obtained from Merck (Darmstadt, Germany).

7.2 Methods

7.2.1 Solid phase peptide synthesis (SPPS)

Bruce R. Merrifield introduced the SPPS methodology in the early 1960's [1], and since then it has become the method of choice for the preparation of native and chemically modified peptides contained up to 50 residues. Also larger polypeptides and small proteins with more than 100 residues have been synthesized by stepwise SPPS [2].

The concept of SPPS is based on the build-up of the peptide backbone from the C-terminal to the N-terminal end, while the C-terminus is anchored to a solid support through a linker (Figure 7.1). The Merrifields chemistry was based on the Boc(*tert*-butyloxycarbonyl) strategy. To avoid the repetitive acidolysis (50% TFA) required in Boc SPPS, new methods were developed. The most successful and extensively explored chemistry for SPPS utilizes the base-labile Fmoc(9-fluorenylmethyloxycarbonyl) group. In the *Fmoc* strategy the initial N^α-*Fmoc* and, when required, side chain protected amino acid is loaded onto the solid support via an acid labile linker [1, 3]. The side chain protecting group are also acid labile, like trityl, Boc, and t-Bu groups (orthogonally protection). After attachment of the N-α-protected C-terminal amino acid to the resin, the temporary N-protection is cleaved with a base (usually 20-40% piperidine in DMF). Then the second N-α-protected and activated (*in situ*) amino acid is coupled to the first one. Several activators may be used [4, 5]. The most widely used coupling method in *Fmoc* SPPS is the active ester method applying HOBt and its corresponding uranium salt analog HBTU. The deprotection/coupling cycles are repeated till the sequence is completed. In the end, the crude product is cleaved from the solid support and simultaneously side-chain deprotected by a concentrated acid solution (TFA) to obtain the peptide as a free acid or amide, depending on the chemical nature of the linker.

The first solid support used by Merrifield was based on polystyrene crosslinked with 1% *m*-divinylbenzene and on a chlorobenzyl linker. This resin is known as the Merrifield resin and has been in the past the standard resin for the synthesis of peptide acids by Boc strategy. However, strong acids like HF are required to remove the peptide chain from the solid support. Since in this work the SPPS has been performed in combination with the *Fmoc* strategy, resin linkers have been employed, which are labile under mild acid conditions (Figure 7.2). In this PhD work the Rink amide MBHA resin and Rink amide-ChemMatrix® resin have been used to prepare the peptide amides. The preloading is not necessary, because the linkers of both Rink amide resins are easily acylated and the resin can be used directly for the automated synthesis.

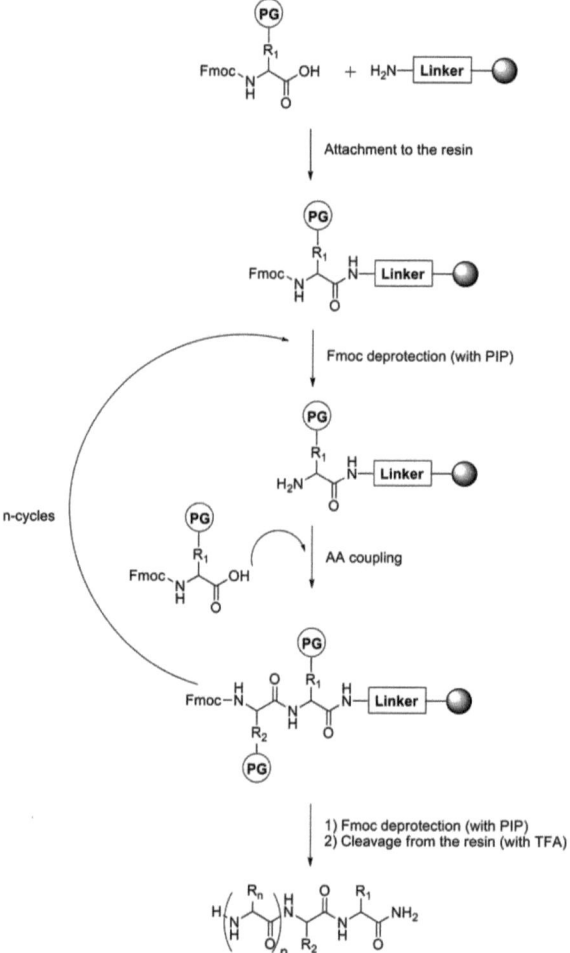

Figure 7.1: Solid phase peptide synthesis of peptide amides employing in *Fmoc* strategy.

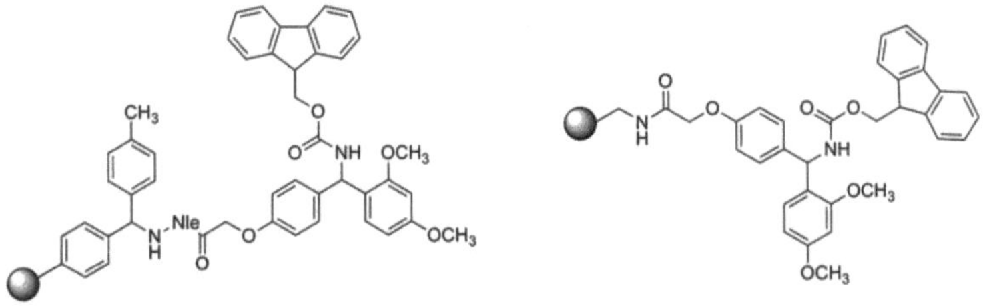

Rink Amide MBHA resin **Rink Amide-ChemMatrix®**

Figure 7.2: Chemical structures of the two resin linkers used in this work for SPPS: Rink amide MBHA resin and Rink amide-ChemMatrix® resin (the Rink amide resins are purchased in the *Fmoc*-protected form).

7.2.2 Circular dichroism (CD) spectroscopy [6]

CD spectroscopy is a very convenient and widely used technique to study the secondary structure of peptides and proteins in solution [7].

In the far UV region, the peptide bond is the main chromophore, from which the content of secondary structure elements such as α-helix, β-sheet, diordered structure, PPII helix, and turns can be estimated (Figure 7.3). The α-helix shows the $n \rightarrow \pi^*$ negative band at 222 nm and the splitted $\pi \rightarrow \pi^*$ negative band at 208 nm and positive one at 193 nm. The β-sheet exhibits a broad negative band near 218 nm (attributed to the $n \rightarrow \pi^*$ transition) and a positive band near 195 nm (attributed to the $\pi \rightarrow \pi*$ transition). Unordered structure and the PPII helix, which is a component of unordered structures, have an intensive negative band near 200 nm (attributed to the $\pi \rightarrow \pi^*$ transition) and the PPII helix has also an additional weak positive band near 225 nm (attributed to the $n \rightarrow \pi^*$ transition).

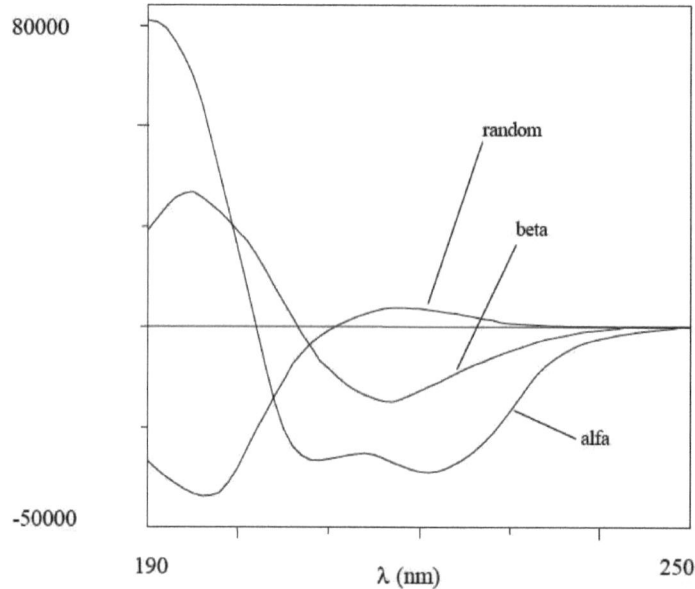

Figure 7.3: CD spectra of poly-L-lysine in aqueous solution at different pH and temperature values: α-helix at pH > 10.5 and R.T., β-structure at pH > 10.5 and T ≥ 50°C and random coil at pH < 10.5 [8].

7.2.3 Steady-state fluorescence spectroscopy

Fluorescence resonance energy transfer (FRET) is a useful tool to study molecular dynamics in biophysical and biochemistry, such as protein-protein interactions, protein-DNA interactions, and protein conformational changes.

Fluorescence resonance energy transfer (FRET) is the non radiative transfer of energy from an electronically-excited donor chromophore (D) to an acceptor chromophore (A) in the ground state through long-range dipole-dipole interactions without emission of a photon [9]. The acceptor absorbs energy at the emission wavelength of the donor, and if fluorescent, emission will be observed when the donor is excited (sensitized emission). Förster proposed in 1948 a theory for long-range molecular interaction by resonance energy transfer, which postulated that the rate of transfer depends on the inverse sixth power of the distance between the donor and the acceptor, which should be in the range of 10 to 100 Å.

7.3 Standard procedure for polypeptide synthesis

7.3.1 Peptide chain assembly by automated SPPS

Peptide chain synthesis was achieved at 0.018 mmol scale in 2 mL vessels applying a *Syro-I* batchwise synthesizer (MultiSynTech, Witten, Germany), and using *Fmoc*-chemistry. The following side chain protecting groups were used: t-Bu for Asp, Glu, Ser, Thr and Tyr; *trityl* for Cys, His, Asn and Gln; *Boc* and 4-Methyltrityl (*Mtt*) for Lys; 2,2,4,6,7-pentamethyl-dihydrobenzofurane-5-sulfonyl (*Pbf*) for Arg. The peptide amides were assembled on Rink amide MBHA resin (loading 0.6 mmol/g) and ChemMatrix Rink amide resin (loading 0.56 mmol/g). Double couplings (40 min each) were performed by in situ activation of N^α-*Fmoc*-amino acids (5 eq) by HBTU/HOBt/DIPEA (5:5:10 eq) in DMF/NMP (80:20, v/v). The Fmoc group was cleaved by 40% piperidine (v/v) in DMF/NMP (80:20, v/v) for 3 min, and then again with 20% piperidine for 10 min.

The N-terminally truncated analogues were acetylated at the N-terminus with acetic anhydride (10 eq) in the presence of DIPEA (10 eq) in DMF for 25 min. Finally, cleavage of the polypeptides from the resin with simultaneously side chain deprotection was achieved with TFA in the presence of 10% (v/v) scavengers (water/TIS/EDT 4:2:4, v/v). After 2h 45 min, the resin was filtered off and the crude products were precipitated with ice-cold diethyl ether and recovered by centrifugation at 3°C for 8 min. Several ether washes/centrifugation cycles were carried out to efficiently remove the scavenger. When Met oxidation occurred, a reductive treatment by TFA/EDT/TMSBr solution (95:3:2, v/v) followed the cleavage of the Id3 polypeptides from the resin. After 1 h, the reduced peptides were precipitated by ice-cold diethyl ether and recovered by centrifugation, as mentioned before. The crude peptides were purified and analyzed by RP-HPLC and mass spectrometry.

7.3.2 Peptide chain elongation by manual SPPS

For the peptide fragments **IV.1** and **IV.2** the last five amino acids (Arg for **IV.1** and Glu for **IV.2**) were coupled manually to the resin-bound Id3 fragments (41-81) (the HLH domain), which were first synthesized by automated solid phase synthesis, as described above, using HBTU (4.8 equiv.), HOBt (5 equiv.) and DIPEA (10 eq.). Double couplings were performed for 45 min each. The success of reactions was controlled by cleaving a small amount of peptide from the resin and analyzing it by analytical HPLC and MALDI-TOF-MS. The peptides were finally cleaved from the resin and simultaneously deprotected with the mixtures $TFA/H_2O/TIS/EDT/Thioanisole$ (88:3:3:4:2, v/v) for 2.5 h. The crude peptides were precipitated from ice-cold diethyl ether and recovered by centrifugation at 3 °C for 7 min. Several ether washes/centrifugation cycles were carried out to efficiently remove the scavengers. The crude peptides were then purified by semi-preparative HPLC and the fractions were analyzed by MALDI-TOF-MS to detect the pure product.

7.3.3 Labeling of the Id3 fragments

The Mtt-group was cleaved from the Lys side chain by 1% TFA in DCM in the presence of 5% TIS [10] in fifteen cycles (1 min each). The free amino group was then acylated with 5-FAM or 5-TAMRA (10 eq) activated with DIC/HOBt (10 eq each) in DMF (600 µL) for 45 min.

The completion of the reaction was checked qualitatively by the ninhydrin test, and quantitatively by performing a small-scale TFA cleavage of the peptide from the resin and by HPLC and MALDI-TOF mass spectrometry analysis.

The two differently labeled peptides were cleaved from the resin with simultaneously side chain deprotection with the mixture of 90% (v/v) TFA and 10% (v/v) scavengers (water/TIS/EDT 2:4:4, v/v) for 3 hours. The crude peptide was precipitated from ice-cold diethyl ether and recovered by centrifugation at 3 °C for 5 min.

7.4 General procedure for peptide purification and characterization

The synthesized peptides were purified by semi-preparative RP-HPLC using the Macherey-Nagel Nucleosil 100-5 C18 column, 5 µm particle size, 250 x 10 mm. The crude peptides were dissolved in 0.1% TFA (v/v) in water and filtered through polytetrafluoroethylene syringe filter. The binary gradient elution system was: (A) TFA (0.06%, v/v) in water and (B) TFA (0.05%, v/v) in ACN. The following gradient was used: 40-65% B over 25 min at a flow rate of 2 mL/min. UV detection at 220 nm was used. The fractions collected were controlled by MALDI-TOF-MS to detect the desired products. After evaporating ACN from the desired fractions, the peptides were lyophilized. Purity of the products after lyophilization was characterized by analytical RP-HPLC using the Macherey-Nagel Nucleosil 100-5 C18 column, 5 µm particle size, 250 x 4 mm. The pure samples were dissolved in 0.1% TFA (v/v) in water. The binary gradient elution system was the same as for the semi-preparative RP-HPLC. The gradients used have been mentioned in the previous chapters. UV detection at 220 nm and 280 nm was used, except for the fluorescence labeled peptides which were additionally detecteted at 480 nm (FAM absorbtion) or 550 nm (TAMRA absorbtion).

All purified peptides were obtained with purity of 80-90% (based on their analytical RP-HPLC profiles).

7.5 Mass spectrometry

Molecular masses were measured on a mass spectrometer from Bruker for MALDI-TOF (matrix-assisted laser desorption time-of-flight) analysis. The samples were dissolved in 0.1% TFA (v/v) in water and mixed 1:2 (v/v) with a solution of α-cyano-4-hydroxycinnamic acid in MeOH/ACN (1:1, v/v).

7.6 UV and CD spectroscopy

All CD spectra were recorded on a Jasco J-815 CD spectropolarimeter using a quartz cell of 0.01 cm path length (Hellma, Mühlheim, Germany). The peptides were dissolved in phosphate buffer (100 mM, pH 7.1), centrifuged and the supernatant was analyzed by UV spectroscopy for the determination of the concentration on the basis of the absorption of the Tyr residues at 280 nm ($\epsilon = 1480$ M^{-1}) on Cary 1E UV-Visible spectrometer from Varian GmbH (Darmstadt, Germany). The UV cuvettes, 0.5 mL with 1 cm path length, were purchased form Hellma. The CD samples were obtained by dilution of freshly prepared stock solutions of the peptides. For each spectrum, two scans were accumulated using a data pitch of 1 nm and a speed scan of 20 nm/min in a range from 190 nm to 260 nm. In the presence of chloride ions the range of wavelength was from 200 nm to 260 nm. The spectrum of buffer was subtracted from that of the peptide to eliminate interferences from cell, solvent, and optical equipment. Noise reduction was obtained by a Fourier transform filter with the program Origin (OriginLab Corporation, Northamton, MA, USA). The ellipticity was expressed as mean-residue molar ellipticity. The obtained dichroic data were evaluated by using DICHROWEB, an online service for the evaluation of CD data [11, 12]. The best fit, characterized by the lowest RMSD value between the calculated and the experimental CD curves was given by the CONTIN algorithm [13, 14].

7.6.1 Procedure for thermal denaturation by CD spectroscopy

The thermal denaturation of peptides was measured by monitoring the ellipticity at 222 nm as a function of the temperature, which was raised at a constant rate of 1 °C/min from 20 to 90 °C, using a temperature-scanning CD system. Renaturation was also recorded from 90 to 20 °C by cooling. The thermal curve of the buffer was subtracted from that of the peptide to eliminate interferences from cell, solvent, and optical equipment.

7.7 Oxidation experiments

7.7.1 Monomer oxidation

The cysteine containing peptides **III.1** and **III.2** (0.7 mg each) were dissolved in 500 μL of degassed 0.1 M ammonium bicarbonate buffer (pH $\sim$ 8) and allowed to oxidize under air atmosphere in an opened Eppendorf vial to produce the disulfide linked dimers **III.1**^{OX} and **III.2**^{OX}. To follow the course of the oxidation reaction, small aliquots were taken immediately and after 3 h, 20 h, and 32 h, acidified by addition of 20% acetic acid and subjected to analytical HPLC and MS measurements.

7.7.2 Disulfide reshuffling and thiol-disulfide exchange measurements

For the disulfide reshuffling the lyophilized **III.1**^{OX} and **III.2**^{OX} ($\sim$ 0.2 mg each) were dissolved in 200 μL 0.1 M ammonium bicarbonate buffer (pH $\sim$ 8) containing 20 mM reduced glutathione and 40 mM oxidized glutathione, under nitrogen atmosphere. After 30 minutes, 1 h, 2 h 30 minutes small aliquots were taken, acidified with 20% acetic acid and subjected to analytical HPLC and MS measurements. In thiol-disulfide exchange experiment the dimer **III.2**^{OX} and the monomer **III.1** were dissolved in 500 μL ammonium bicarbonate buffer (pH $\sim$ 8), under nitrogen atmosphere. After 30 minutes, 1 h, 2 h 30 minutes small aliquots were taken, acidified with 20% acetic acid and subjected to analytical HPLC and MS measurements.

7.8 Procedure for fluorescence spectroscopy analysis

7.8.1 Fluorescence spectroscopy of the labeled Id3 peptides

The labeled peptides were dissolved in 100 mM phosphate buffer (pH 7.1) and the peptide concentration was determined by the UV absorbance at 205 nm [15] by using the following equation:

$$c(mg/ml) = \frac{A_{205nm}}{31} \qquad (7.1)$$

The 5 μM samples solutions were obtained by dilution of the stock solutions previously prepared. The fluorescence emission spectra were recorded at R.T. on a Varian Cary Eclipse fluorimeter by excitation at 488 nm (λ_{exc} of FAM) or at 550 nm (λ_{exc} of TAMRA).

7.8.2 Titration of partially folded Id3 HLH with folded Id3 HLH

The labeled peptide **V.2** (partially folded Id3 HLH) and the unlabeled peptide **V.8** (fully folded Id3 HLH) were dissolved in 100 mM phosphate buffer (pH 7.1) and the peptide concentration was determined by the UV absorbance at 205 nm as previously explained.

The 5 μM sample of peptide **V.2** was obtained by dilution of the stock solution previously prepared, and a stock solution of unlabeled peptide **V.8** was used to titrate the 5 μM solution of the labeled peptide. The emission spectra after each addition were recorded after 5 minutes of incubation at room temperature by excitation at 488 nm (λ_{exc} of FAM). The spectra were then corrected by the dilution factor.

References

[1] Fields, G. B.; Noble, R. L., Solid phase peptide synthesis utilizing 9-fluorenylmethoxycarbonyl amino acids. *Int J Pept Protein Res* **1990**, 35 (3), 161-214.

[2] Svobodova, J.; Cabrele, C., Stepwise solid-phase synthesis and spontaneous homodimerization of the helix-loop-helix protein Id3. *Chembiochem* **2006**, 7 (8), 1164-8.

[3] Sheppard, R., The fluorenylmethoxycarbonyl group in solid phase synthesis. *J Pept Sci* **2003**, 9 (9), 545-52.

[4] Albericio, F., Developments in peptide and amide synthesis. *Curr Opin Chem Biol* **2004**, 8 (3), 211-21.

[5] Marder, O.; Albericio, F., Industrial application of coupling reagents in peptide synthesis. *Chimica Oggi* **2003**, 21, 6-11.

[6] Kelly, S. M.; Price, N. C., The use of circular dichroism in the investigation of protein structure and function. *Curr Protein Pept Sci* **2000**, 1 (4), 349-84.

[7] Greenfield, N. J., Methods to estimate the conformation of proteins and polypeptides from circular dichroism data. *Anal Biochem* **1996**, 235 (1), 1-10.

[8] Greenfield, N.; Fasman, G. D., Computed circular dichroism spectra for the evaluation of protein conformation. *Biochemistry* **1969**, 8 (10), 4108-16.

[9] Sapsford, K. E.; Berti, L.; Medintz, I. L., Materials for fluorescence resonance energy transfer analysis: beyond traditional donor-acceptor combinations. *Angew Chem Int Ed Engl* **2006**, 45 (28), 4562-89.

[10] Hoogerhout, P.; Stittelaar, K. J.; Brugghe, H. F.; Timmermans, J. A.; ten Hove, G. J.; Jiskoot, W.; Hoekman, J. H.; Roholl, P. J., Solid-phase synthesis and application of double-fluorescent-labeled lipopeptides, containing a CTL-epitope from the measles fusion protein. *J Pept Res* **1999**, 54 (5), 436-43.

[11] Whitmore, L.; Wallace, B. A., DICHROWEB, an online server for protein secondary structure analyses from circular dichroism spectroscopic data. *Nucleic Acids Res* **2004**, 32 (Web Server issue), W668-73.

[12] Lobley, A.; Whitmore, L.; Wallace, B. A., DICHROWEB: an interactive website for the analysis of protein secondary structure from circular dichroism spectra. *Bioinformatics* **2002**, 18 (1), 211-2.

[13] Provencher, S. W.; Glockner, J., Estimation of globular protein secondary structure from circular dichroism. *Biochemistry* **1981**, 20 (1), 33-7.

[14] van Stokkum, I. H.; Spoelder, H. J.; Bloemendal, M.; van Grondelle, R.; Groen, F. C., Estimation of protein secondary structure and error analysis from circular dichroism spectra. *Anal Biochem* **1990**, 191 (1), 110-8.

[15] Tombs, M. P.; Souter, F.; Maclagan, N. F., The spectrophotometric determination of protein at 210 millimicrons. *Biochem J* **1959**, 73, 167-71.

Summary

The Id3 protein is a member of the family of the Id proteins, which act as inhibitors of DNA binding and cell differentiation (Id1-4) and represent the class V of the large family of the helix-loop-helix (HLH) transcription factors. The Id3 protein displays a 119-long amino acid sequence, the HLH region spans residues 41-81, with helix-1 41-56, the loop 57-65 and helix-2 66-81. Together with the other Id proteins, Id3 is involved in a wide range of biological events by forming heterodimers with parent basic-HLH transcription factors. The resulting dimers are no more able to bind the DNA and, as a consequence, basic-HLH mediated expression of specific genes is blocked. The Id proteins are potential targets in cancer therapy, as they are upregulated in several types of tumors, promoting tumor growth and metastasis.

In this PhD work three different approaches were used to study and control the self-association of the Id3 HLH domain, focusing both on the oligomerization order and topology of the helix packing. In addition, it was investigated, how environmental changes affect the folding behavior of the HLH domain of Id3 and its parent protein Id4.

The Id3 protein fragments required for the studies were synthesized by solid-phase methodology using the Fmoc-chemistry. Purification of the crude products was carried out by semi-preparative HPLC. Identification and characterization of the pure peptides was based on MALDI-TOF-MS and analytical HPLC. Structural investigation was based on the application of the CD and fluorescence spectroscopy.

The first approach used in this work is based on the synthesis of a disulfide-bonded Id3 HLH domain, as previously done by S. Kiewitz with the parent Id1 HLH domain in the Cabrele group. Two analogs of the Id3 HLH peptide 41-81 were prepared and subjected to air-oxidation towards the formation of two well-structured covalent homodimers. The latter were differing from each other in the location of the disulfide bridge (N-N linkage or C-C-linkage) and in the presence/absence of a terminal G_3 motif. This allowed to perform thiol/disulfide exchange studies and to detect the formation of possible mixed dimers upon reshuffling conditions by using mass spectrometry. The air-oxidation of both analogs proceeded easily and almost to completeness, which suggests that the formation of the homodimers was favorable in both cases, and that a parallel orientation of the first and second helices, respectively, is favored in the HLH dimer. As the thiol/disulfide exchange assay did not show the formation of covalent heterodimers, it can be assumed that an interchain antiparallel arragement of the HLH domains is unfavorable. Morover, the conversion of the homodimer containing the N-N linkage into mixed disulfides with glutathione was found to be easier than that of the homodimer containing the C-C linkage. This may reflect the superior intrinsic helix propensity of the C-terminal helix-2 with respect to the N-terminal helix-1, which could then lead to a tight self-packing of the helix-2 compared to a moderate or weak self-packing of helix-1. Altogether these results support a preferred parallel orientation of the two HLH domains in the dimer, as it was also found in the Id1 HLH dimer by S. Kiewitz and C. Cabrele and in the recently published NMR solution and crystal structures of the Id3 and Id2 HLH regions.

The second approach applied to control the self-association of the Id3 HLH domain towards homodimers was based on electrostatic interactions between an Id3 HLH analog containing an additional positively charged N-tail (R_5) and another one containing a negatively charged N-tail (E_5). The conformational characterization of the two analogs by CD spectroscopy has revealed that the presence of the positively charged region at the N-terminus of the Id3 HLH motif has a negative effect on the HLH folding. In contrast, the presence of the negatively charged region further stabilized the HLH folding.

The results of the hetero-association study between the two analogs suggest that they interact with each other and form higher-order oligomers. The thermal unfolding/refolding studies on the two analogs confirm the positive effect of the N-terminal negative charges and the negative effect of the N-terminal positive charges on the HLH fold.

The last approach was based on FRET studies. For this purpose new fluorescent Id3 fragments were synthesized. We chose FAM/TAMRA as FRET pair because both can be easily and selectively introduced in the peptide chain. The effect of the labeling on the conformation was first controlled by CD spectroscopy. For example, the CD spectra of labeled and native analogs of the sequence 48-92 of Id3 showed that the positions 76 and 84 are tolerant for labeling, although the presence of TAMRA in the position 84 induced partial peptide aggregation. Similarly, the labeling of position 61 in the sequence 31-61 of Id3 was well tolerated, but the TAMRA label induced peptide aggregation also in this case. This is in agreement with the observation that the fluorescence emission of the TAMRA peptides was very weak, which could be explained by self-quenching. Interestingly, the labeling with FAM and TAMRA in position 76 induced a partial transition from α-helix to 3_{10}-helix. FRET studies between the FAM- and TAMRA-labeled sequences 48-92 of Id3 at position 76 revealed the presence of weak interactions between the two peptides. Similar results were found in the case of the interaction between the peptide labeled with FAM in position 84 and the peptide labeled with TAMRA in position 76. A further fluorescence titration experiment revealed the ability of the partially folded HLH domain to interact with the fully folded HLH domain in a 1:2 ratio.

Finally, the behavior of the Id3 and Id4 HLH domains upon environmental changes was investigated by CD spectroscopy. In this work we showed how the salts chol-dhp (choline dihydrogen phosphate) and chol-Cl (choline chloride) affected the Id HLH fold in different manner. However, the changes were related to the target peptide too. Indeed, chol-dhp stabilized the helix-rich fold of the Id4 HLH, while destabilizing the fold of the Id3 HLH. In contrast, chol-Cl destabilized the fold of the Id4 HLH, while stabilizing it for the Id3 HLH. The different behavior between the two HLH domains was attributed to the presence of different amino acid patterns at the polar surface of helix-1.

In conclusion, in the presented PhD work Id3 polypeptide fragments as well as their fluorescence labeled analogs were successfully prepared by the SPPS methodology using Fmoc chemistry. Additional information on the self-association and folding behavior of the Id3 HLH domain was provided: in particular, it appears that the helix-2 fragment might act as a locally ordered structure in the early states of the folding process. This might then trigger the oligomerization of the Id3 protein by helix-2 self-interactions, possibly favoring successive contacts between the helix-2 and the nascent helix-1, thus resulting in the parallel four-helix bundle. It could be also shown that a basic motif adjacent to the N-terminus of the Id3 HLH domain triggers self-aggregation, contrarily to an acidic motif. Further, the study of the effect of cosolutes on the HLH fold of Id3 and Id4 suggests, that non-conserved residues, in particular in the helix-1, are likely to play a role in the interaction with the environment, thus determining a different behavior of the two Id HLH sequences in molecular recognition events, despite their high sequence identity.

Appendix: MS and HPLC data

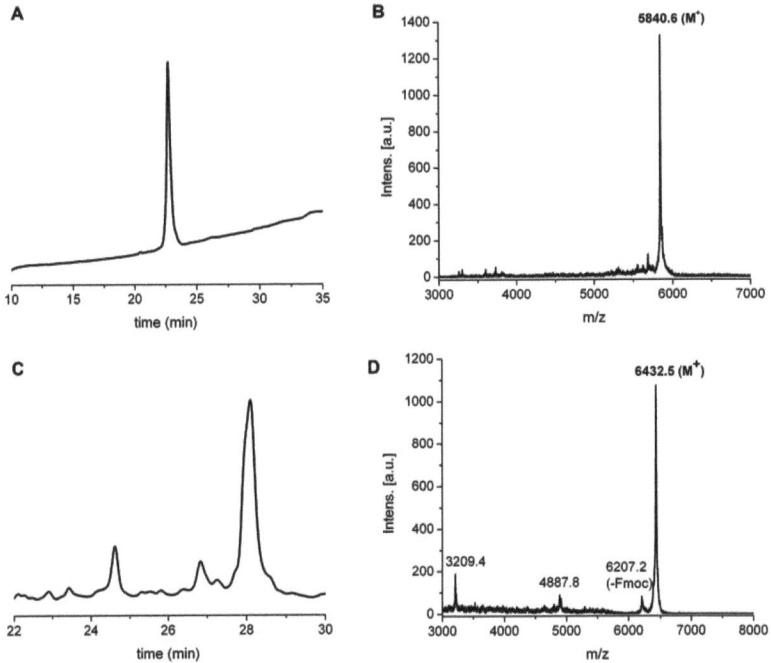

Figure A.1: HPLC profiles of the peptides **II.2** (**A**) and **II.3b** (**C**) (gradient: 20% ACN for 5 min., 20-80% ACN over 40 min.). MALDI spectra of the peptides **II.2** (**B**) and **II.3b** (**D**).

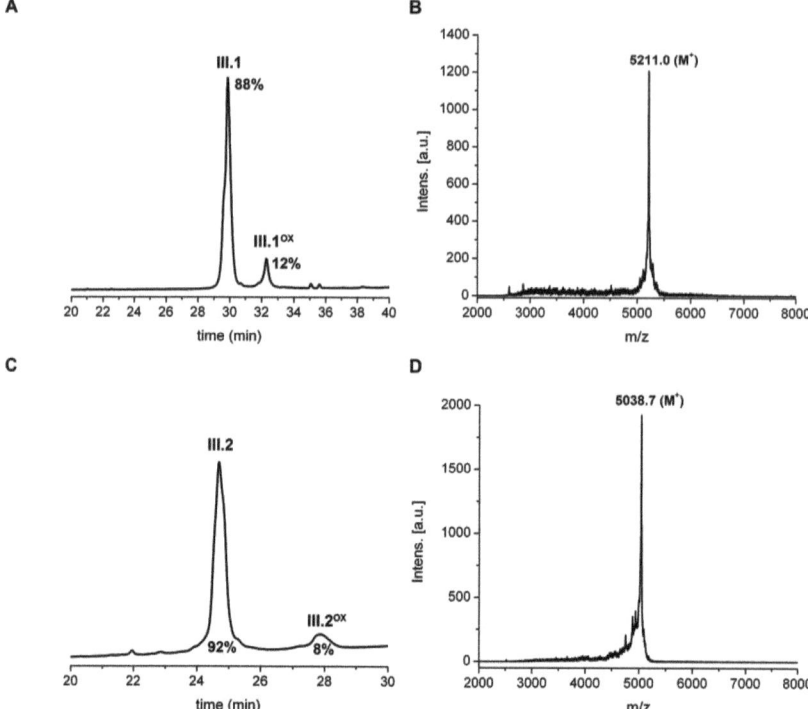

Figure A.2: HPLC profiles of the peptides **III.1** (**A**, gradient: 30% ACN for 5 min., 30-70% ACN over 30 min.) and **III.2** (**C**, gradient: 30% ACN for 5 min., 30-70% ACN over 30 min.). MALDI spectra of the peptides **III.1** (**B**) and **III.2** (**D**).

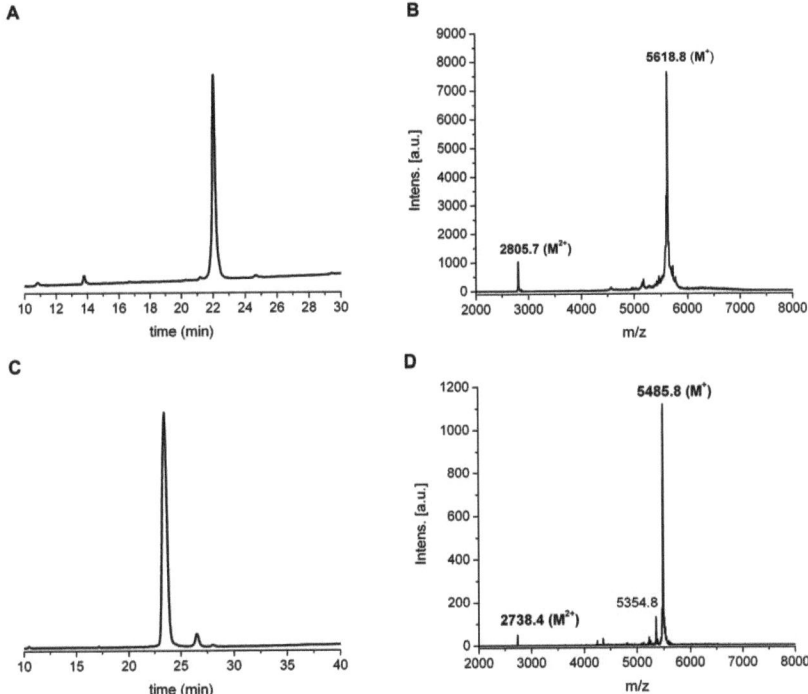

Figure A.3: HPLC profiles of the peptides **IV.1** (**A**, gradient: 30% ACN for 5 min., 30-80% ACN over 35 min.) and **IV.2** (**C**, gradient: 30% ACN for 5 min., 30-80% ACN over 35 min.). MALDI spectra of the peptides **IV.1** (**B**) and **IV.2** (**D**).

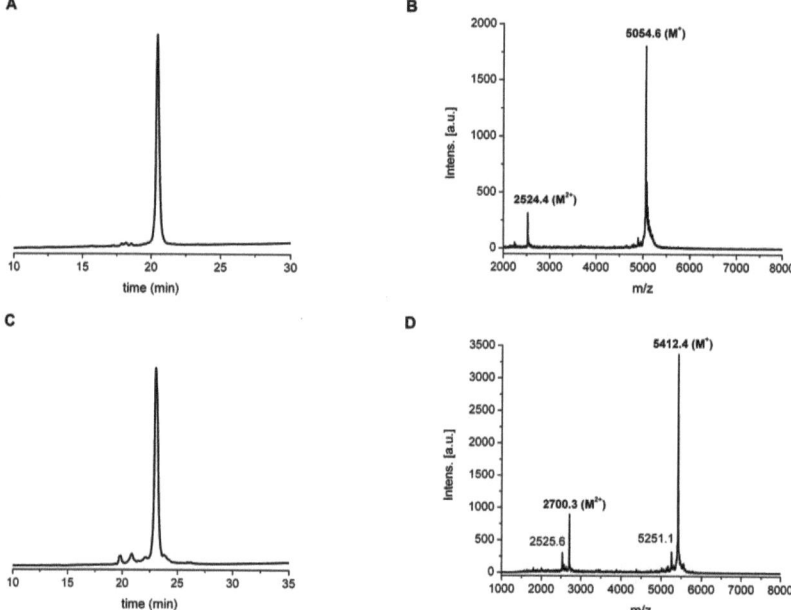

Figure A.4: HPLC profiles of the peptides **V.1** (**A**, gradient: 40% ACN for 5 min., 40-85% ACN over 40 min.) and **V.2** (**C**, gradient: 40% ACN for 5 min., 40-85% ACN over 40 min.). MALDI spectra of the peptides **V.1** (**B**) and **V.2** (**D**).

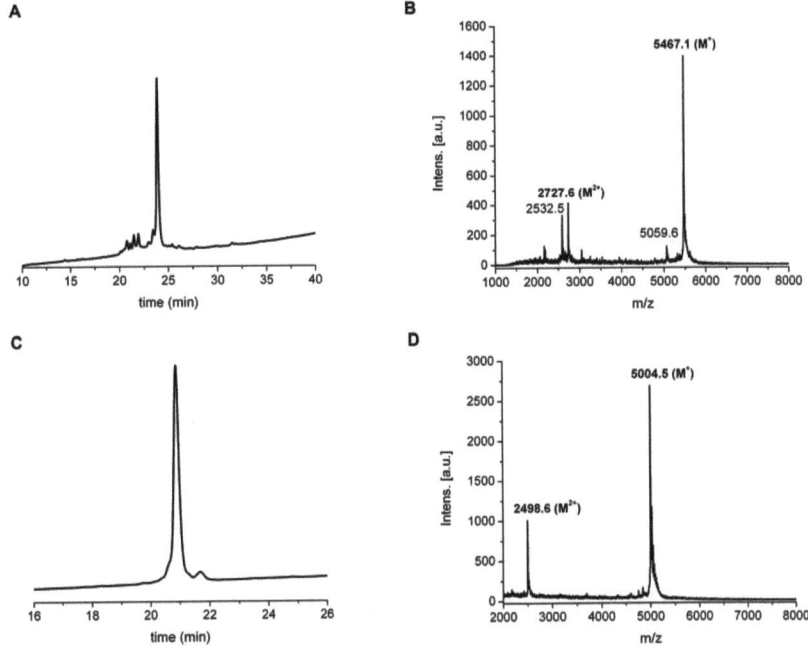

Figure A.5: HPLC profiles of the peptides **V.3** (**A**, gradient: 40% ACN for 5 min., 40-85% ACN over 40 min.) and **V.4** (**C**, gradient: 40% ACN for 5 min., 40-85% ACN over 40 min.). MALDI spectra of the peptides **V.3** (**B**) and **V.4** (**D**).

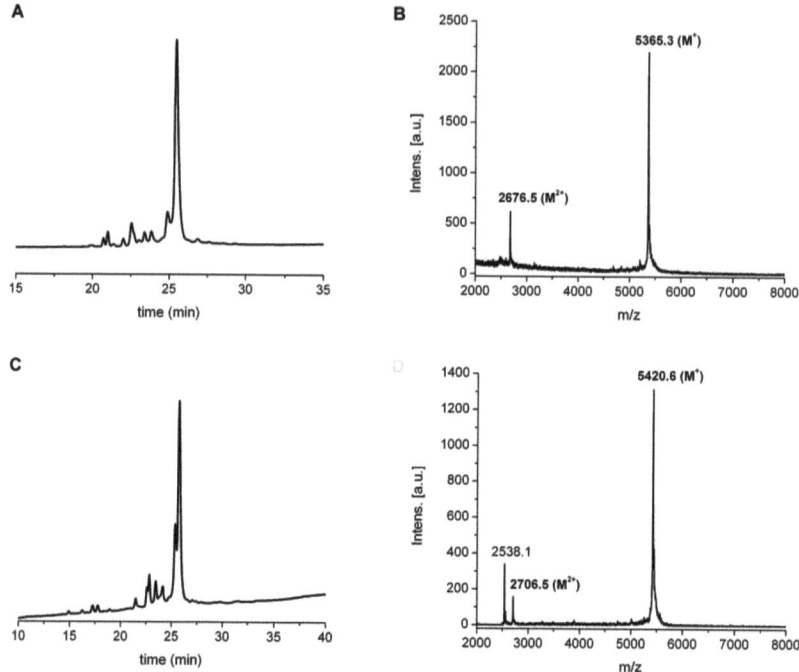

Figure A.6: HPLC profiles of the peptides **V.5** (**A**, gradient: 40% ACN for 5 min., 40-85% ACN over 40 min.) and **V.6** (**C**, gradient: 40% ACN for 5 min., 40-85% ACN over 40 min.). MALDI spectra of the peptides **V.5** (**B**) and **V.6** (**D**).

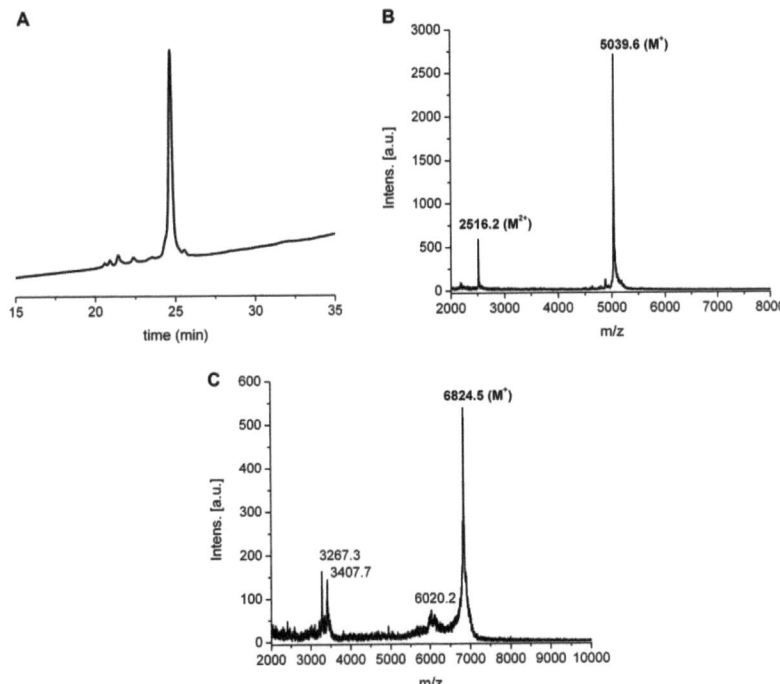

Figure A.7: HPLC profiles of the pure peptides **V.7** (**A**, gradient: 40% ACN for 5 min., 40-85% ACN over 40 min.). MALDI spectra of the peptides **V.7** (**B**) and **V.8** (**C**).

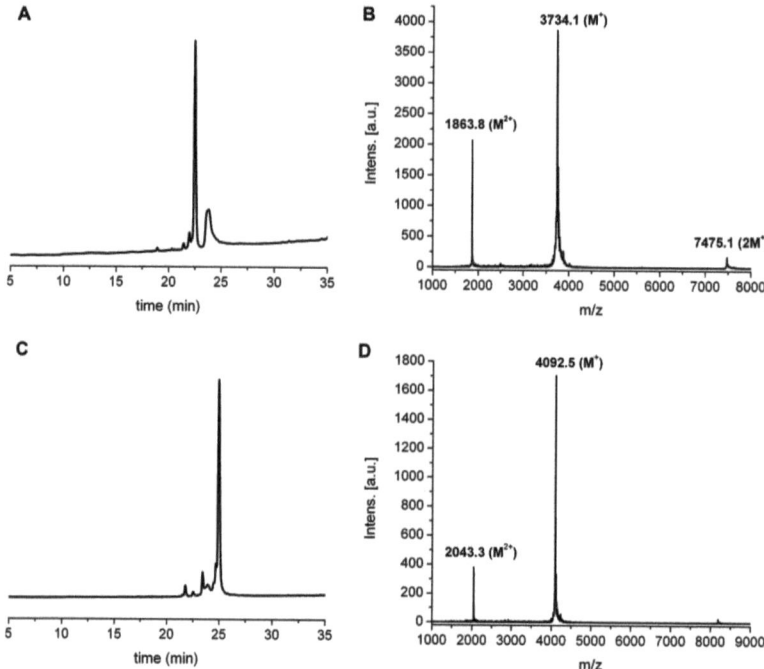

Figure A.8: HPLC profiles of the pure peptides **V.9** (**A**, gradient: 20% ACN for 5 min., 20-80% ACN over 40 min.) and **V.10** (**C**, gradient: 20% ACN for 5 min., 20-80% ACN over 40 min.). MALDI spectra of the pure peptides **V.9** (**B**) and **V.10** (**D**).

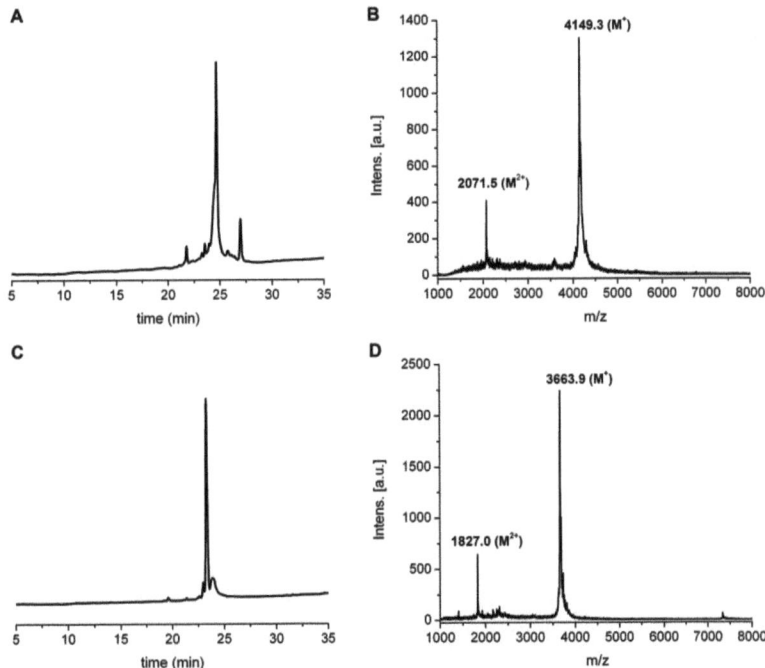

Figure A.9: HPLC profiles of the peptides **V.11** (**A**, gradient: 20% ACN for 5 min., 20-80% ACN over 40 min.) and **V.12** (**C**, gradient: 20% ACN for 5 min., 20-80% ACN over 40 min.). MALDI spectra of the peptides **V.11** (**B**) and **V.12** (**D**).

Acknowledgements

We are at the last page of this thesis, this is the place to say thanks to the people who in these years helped and supported me in different manners. First of all I would like to thank **Prof. Dr. Chiara Cabrele**, who gave the opportunity to perform my PhD at RUB in Bochum and the chance to work in her group. I am specially thankful to her because she gave me an interesting research project in the great world of peptides and always supported the work. I have learned a lot from her and I think she is really able to motivate people.

I am grateful to **Prof. Dr. Günter von Kiedrowski**, **Dr. Wolf Matthias Pankau** and their group who made the use of their equipment and their precious help available; to **Mr. Rolf Breuckmann** for the technical help with the MALDI instrument; to **Florian Kaschuba** concerning the IT for making the web page of the Cabrele's working group. Moreover, I would like to thank **Miss Stefanie Wittmann** for her help and support in the bureaucratic issues.

Besides I thank the working group of **Prof. Dr. Christian Herrmann** for the use of the circular dichroism spectrometer.

Moreover, I would like to thank all the people from the lab who strongly contributed to my experience in Bochum, in particular my PhD colleague **Saskia Nuekirche** for the nice time in the lab and for the precious help with the German language. Then another special thanks to a colleague and good friend **Michele Zuliani**, for his sympathy, the nice discussions and the nice time spent together during parties. And I would like to thank **Jorge Malaver**, one of my best friend here, for his sympathy, the nice time spent together during lunch breaks and outside the university, and precious help when he taught me Spanish.

Furtheremore, I would like to thank the people who supported me from the distance, in particular a colleague and good friend **Dr. Andrea Caporale** for his help in the corrections of my thesis and to be always willing to help and support me.

At the end I am especially grateful to my parents, **Teresa** and **Giuseppe**, together with my brother **Enrico**, for their constant support, they always gave me the best and they always encouraged me. I know that they are proud of me and this makes me happy.

Printed by Books on Demand GmbH, Norderstedt / Germany